Applied Thermodynamics

Collection of Formulas

HANS HAVTUN

Art. No. 38856
ISBN 978-91-44-10577-2
Edition 1:1

© The Author and Studentlitteratur 2014

www.studentlitteratur.se
Studentlitteratur AB, Lund

Cover design by: Pernilla Rosander

Printed by Dimograf, Poland 2014

Preface

This Collection of Formulas is adapted to the textbook *Tillämpad termodynamik*[1] (Applied Thermodynamics) by Ingvar Ekroth and Eric Granryd. The textbook is used as literature in undergraduate courses in Thermodynamics during the first years of University level programs in Energy and Mechanical Engineering.

Although the textbook is in Swedish, this Collection of Formulas has been written in English to provide the students with necessary bilingual skills in Thermodynamics. Even though mixing two languages may hinder the learning initially, I am of the firm belief that this will eventually be beneficial for the students.

The purpose with this book is to be a convenient helping aid when solving problems in Thermodynamics. Suitable problems adapted to the textbook can be found in the workbook *Arbetsmaterial till Tillämpad termodynamik*[2]. This Collection of Formulas could also be used as a helping aid during examination.

Stockholm, September 2014

Hans Havtun

---

[1] I. Ekroth , E. Granryd, *Tillämpad termodynamik*, Studentlitteratur, Lund, 2006, ISBN 978-91-44-03980-0

[2] H. Havtun, *Arbetsmaterial till Tillämpad termodynamik*, Studentlitteratur, Lund, 2014, ISBN 978-91-44-10576-5

Table of Contents

Preface   iii
Nomenclature   vii
1. Concepts and Definitions
    1.1   Systems   1
    1.2   Properties   1
    1.3   Temperature Scales   1
    1.4   Common Relations   1
    1.5   Hydrostatic Pressure   2
    1.6   Archimedes' Principle   2

2. First Law of Thermodynamics
    2.1   Work   3
    2.2   Heat   3
    2.3   First Law of Thermodynamics   3
    2.4   Energy Balance for a Closed System   3
    2.5   Energy Balance for an Open System   4
    2.6   Energy Analysis of Cycles   5

3. Ideal Gas Model
    3.1   Ideal Gas Law   7
    3.2   Mixtures of Ideal Gases   7
    3.3   Specific Heat   8
    3.4   Kinetic Gas Theory   10

4. Processes with Ideal Gases
    4.1   Simple Processes   11
          4.1.1   Isothermal Process   11
          4.1.2   Isochoric Process   11
          4.1.3   Isobaric Process   11
          4.1.4   Isentropic Process   12
          4.1.5   Polytropic Process   12
    4.2   Carnot Cycle   13
    4.3   Cycle Definitions   14

5. Second Law of Thermodynamics, Entropy and Exergy
    5.1   Second Law of Thermodynamics   15
    5.2   Entropy   15
    5.3   Carnot Cycle in an s–T-diagram   16
    5.4   Exergy   16

6. Gas Power Cycles
    6.1   Otto Cycle   17
    6.2   Diesel Cycle   17
    6.3   Joule/Brayton Cycle   17
    6.4   Ericsson Cycle   18
    6.5   Stirling Cycle   18

7. Thermodynamic Relations for Simple Compressible Substances

7.1	p–v–T-Surface	19
7.2	Two-phase Areas	19
7.3	Generalized Compressibility Chart	20

8. Vapor Power Systems

8.1	Rankine Process	21
8.2	Feedwater Heating	22
8.3	Reheating	22

9. Refrigeration and Heat Pump Systems

9.1	Basic Vapor Compression Cycle	23
9.2	Vapor Compression Cycle with Superheat and Subcooling	24

10. Fluid Mechanics

10.1	Incompressible Fluid Mechanics	25
10.2	Friction Factors	26
10.3	Loss Coefficients	26
10.4	Compressible Fluid Mechanics	27
10.5	Hugoniot Equation	27
10.6	Velocity in a Nozzle	27
10.7	Critical Pressure Ratio	28
10.8	Simple Nozzles	28
10.9	De Laval Nozzles	28
10.10	Media Flow of an Ideal Gas in a Nozzle	28
10.11	Design of De Laval Nozzles for Ideal Gases	29
10.12	Expansion Processes in Nozzles	29

11. Heat Transfer

11.1	Conduction	31
11.2	Convection	31
	11.2.1 Natural Convection	32
	11.2.2 Forced Convection	33
11.3	Radiation	34
11.4	Heat Exchangers	35
11.5	Heat Transfer through Walls	36

12. Psychrometrics

12.1	Introduction	37
12.2	Steady State Mixing of Two Streams of Humid Air	38
12.3	Mass Transfer Influence on Heat Transfer	39
12.4	Lewis Relation	39
12.5	Bäckström Relation	40
12.6	Total Heat Transfer	40

Tables and Diagrams   41

Nomenclature

Symbol		Introduced in section
a	velocity of sound, [m/s]	10.4
A	cross-sectional area, [m²]	10.1
$|b|$	friction work per unit mass, [Nm/kg]	2.4
c_p	specific heat at constant pressure, [J/(kg·K)]	3.3
c_v	specific heat at constant volume, [J/(kg·K)]	3.3
C	number of components in the system (Gibb's phase rule)	1.2
C_B	constant in Bäckström's relation, [K/bar]	12.5
COP_1	coefficient of heating performance	2.6
COP_{1C}	Carnot coefficient of heating performance	4.2
COP_2	coefficient of cooling performance	2.6
COP_{2C}	Carnot coefficient of cooling performance	4.2
C_{Ra}	temperature-dependent function, "Rayleigh function", [1/(m³·K)]	11.2.1
d	pipe diameter, [m]	10.1
e	exergy (availability), [Nm/kg]	5.4
E	work, [Nm]	1.4
$\dot{E}$	net rate of work transfer, [W]	1.4
$\dot{E}_c$	net rate of work transfer for a cycle, [W]	4.3
f	number of degrees of freedom for energy	3.4
f_1	friction factor	10.1
F	number of properties to be specified in Gibb's phase rule, [-]	1.2
g	gravity acceleration constant (= 9.81 m/s²)	2.5
Gr	Grashof number	11.2.1
Gz	Graetz number	11.2.2
h	specific enthalpy, [J/kg]	2.2
H	wall height, [m]	11.2.1
k	overall heat transfer coefficient, [W/(m²·K)]	11.4
k_B	Boltzmann's constant (= $1.38062 \cdot 10^{-23}$ J/K)	3.4
L	pipe length, [m]	10.1
L_t	thermal entry length, [m]	11.2.2
m	mass of system, [kg]	1.4
m_m	mass of one molecule, [kg]	3.4
$\dot{m}$	mass flow rate, [kg/s]	1.4
M	molecular weight, [kg/kmol]	3.1
M_m	molecular weight of mixture, [kg/kmol]	3.2
Ma	Mach number	10.4

Symbol		Introduced in section
n	polytropic exponent	4.1.5
p	pressure, [Pa]	2.1
p_k	critical pressure, [Pa]	7.3
p_r	reduced pressure	7.3
p^*	critical pressure (compressible flow), [Pa]	10.7
Ph	number of phases in the system (Gibb´s phase rule)	1.2
Pr	Prandtl number	11.2.1
q	heat per unit mass, [J/kg]	1.4
q_r	total heat (including friction work) per unit mass, [J/kg]	2.4
Q	heat, [J]	1.4
$\dot{Q}$	heat flow (net rate of heat transfer), [W]	1.4
r	latent heat of vaporization, [J/kg]	7.2
R	individual gas constant, [J/(kg·K)]	3.1
Ra	Rayleigh number	11.2.1
R_M	universal gas constant (= 8314.3 J/(kmol·K))	3.1
Re	Reynolds number	10.1
s	entropy, [J/(kg·K)]	5.1
t	temperature on the Celsius scale, [°C]	1.3
t_{dew}	dew point, [°C]	12.1
T	temperature on the absolute scale (Kelvin scale), [K]	1.3
T_k	critical temperature, [K]	7.3
T_r	reduced temperature	7.3
u	internal energy per unit mass, [J/kg]	2.2
v	specific volume, [m³/kg]	2.1
v_{ri}	pseudoreduced specific volume	7.3
$\dot{V}$	volumetric flow rate, [m³/s]	1.4
V_s	swept volume [m³]	4.3
w	velocity, [m/s]	2.5
x	vapor quality	7.2
x_w	water content, [kg H_2O/kg dry air]	12.1
z	height above an arbitrarily chosen reference level, [m]	2.5
z_r	compressibility factor	7.3

Greek

α_d	diffusion heat transfer coefficient, [W/(m²·K)]	12.5
α_k	convection heat transfer coefficient, [W/(m²·K)]	11.2
α_s	radiation heat transfer coefficient, [W/(m²·K)]	11.3

Symbol		Introduced in section
β	volume expansion coefficient, [1/K]	11.2.1
δ	wall thickness, [m]	11.1
Δp_f	pressure drop, [Pa]	10.1
ε	emissivity	11.3
ε	work per unit mass, [Nm/kg]	1.4
ε_c	net work per unit mass from a cycle, [Nm/kg]	2.6
ε_t	technical work per unit mass (work for open system), [Nm/kg]	2.5
ε_{tr}	reversible technical work per unit mass, [Nm/kg]	2.5
ε_y	volume change work per unit mass, [Nm/kg]	2.4
ε_{yr}	reversible volume change work per unit mass, [Nm/kg]	2.1
φ	relative humidity	12.1
η_1	temperature efficiency	11.4
η_2	temperature efficiency	11.4
η_{Cd}	Carnot efficiency of refrigerant	9.1
η_t	thermal efficiency	2.6
η_{tC}	Carnot thermal efficiency	4.2
η_K	isentropic compressor efficiency	6.3
η_T	isentropic turbine efficiency	6.3
κ	specific heat ratio (isentropic exponent)	3.3
λ	thermal conductivity, [W/(m·K)]	11.1
μ	dynamic viscosity, [Pa·s]	10.1
ν	volumetric fraction	3.2
ν	kinematic viscosity, [m²/s]	10.1
ϑ	temperature difference, [K]	11.1
ϑ_m	logarithmic mean temperature difference, [K]	11.4
ρ	density, [kg/m³]	10.1
σ	Stefan-Boltzmann constant (= $5.67 \cdot 10^{-8}$ W/(m²·K⁴))	11.3
τ	time, [s]	2.5
υ	compression ratio	6.1
ξ	mass fraction	3.2
ψ	mole fraction	3.2
ζ	loss coefficient	10.1

1. Concepts and Definitions

1.1 Systems

In order to perform thermodynamic analysis, we need to define a *system*. The system can be whatever we want to analyze, for instance a heat exchanger or an entire nuclear power plant. A system is distinguished from its surroundings by the *system boundary*. In general, two different types of systems are distinguished: *open* and *closed* systems. The difference between the two types is that the *open system allows mass transfer across the system boundary* to occur, while the closed system does not.

1.2 Properties

Properties are used to describe the *state* of a system. Their values are independent of how the state has been achieved i.e. independent of the history of the system. The number of properties to be specified, F, to define the state of a system is given by *Gibbs' phase rule*

$$F = C - Ph + 2 \qquad\qquad [1.1]$$

where   C   : number of components in the system,
            Ph   : number of phases in the system.

1.3 Temperature Scales

Relation between the temperature on the Celsius scale, $t(°C)$, and the Kelvin scale $T(K)$

$$T(K) = t(°C) + 273.15 \qquad\qquad [1.2]$$

Relation between the Celsius scale, $t(°C)$, and the Fahrenheit scale, $t(°F)$

$$t(°F) = 1.8 \cdot t(°C) + 32 \qquad\qquad [1.3]$$

1.4 Common Relations

$\dot{m}$: mass flow rate,
$\dot{V}$: volumetric flow rate,

Q : heat,
$\dot{Q}$: heat flow (net rate of heat transfer).

$$q = \frac{Q}{m} = \frac{\dot{Q}}{\dot{m}} \qquad\qquad [1.4]$$

E : work (Nm),
$\dot{E}$: net rate of work transfer.

$$\varepsilon = \frac{E}{m} = \frac{\dot{E}}{\dot{m}} \qquad\qquad [1.5]$$

1.5 Hydrostatic Pressure

The *influence of elevation on the atmospheric pressure* can be deduced by balancing the vertical forces on a media element. For a small change in elevation, dz, the change in atmospheric pressure, dp, can be found with:

$$dp = -\rho \cdot g \cdot dz \qquad [1.6]$$

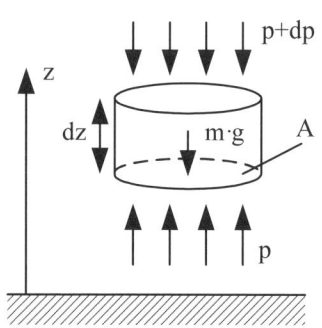

As the density of the air is dependent on the elevation, integrating equation 1.6 requires a relation between the density and the elevation, z.

$$\int_{p_{z=0}}^{p} dp = -g \cdot \int_{0}^{z} \rho \cdot dz \qquad [1.7]$$

A related topic is how the (static) *pressure changes with the depth below a liquid surface.* This dependence can be deduced in a similar manner as for the atmospheric pressure.

For a small change in depth, dz, the change in the pressure, dp, can be found with:

$$dp = \rho \cdot g \cdot dz \qquad [1.8]$$

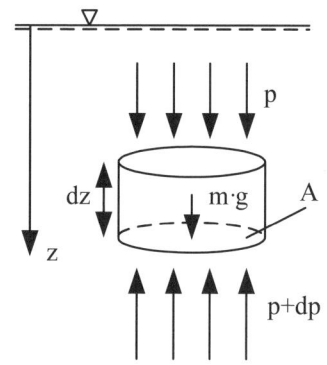

For liquids, the density can usually be assumed constant, the integration is hence greatly simplified:

$$\int_{p_{z=0}}^{p} dp = \rho \cdot g \cdot \int_{0}^{z} dz \qquad [1.9]$$

After integration, we find the final expression:

$$p = p_{z=0} + \rho \cdot g \cdot z \qquad [1.10]$$

where: $p_{z=0}$: pressure at the liquid surface (z = 0),
g : gravity acceleration constant (= 9.81 m/s²),
z : height or depth,
ρ : density.

1.6 Archimedes' Principle

The buoyancy force of a body with submerged volume V in an ambient fluid can be found by:

$$F = g \cdot \left(V \cdot \rho_{amb} - m_{body} \right) \qquad [1.11]$$

For bodies totally submerged in the ambient fluid, equation 1.11 can be reformulated:

$$F = g \cdot V \cdot \left(\rho_{amb} - \rho_{body} \right) \qquad [1.12]$$

2 First Law of Thermodynamics

In this section, α and β denote the state before and after the process respectively.

2.1 Work

The specific work needed to change the volume of a closed system, ε_{yr}, is given by

$$\varepsilon_{yr} = \int_{\alpha}^{\beta} p \cdot dv \qquad [2.1]$$

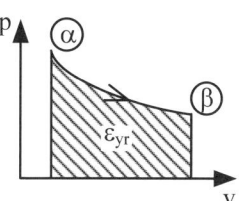

or, in differential form

$$d\varepsilon_{yr} = p \cdot dv \qquad [2.2]$$

where p : pressure in the system,
          v : specific volume ($v = V/m = 1/\rho$).

As can be seen, the specific work, ε_{yr}, is dependent on the function $p = f(v)$, hence *ε_{yr} is not a property*.

2.2 Heat

<u>Definition</u>: *Heat*, q, is the energy exchange between two bodies caused by a temperature difference.

<u>Definition</u>: *Internal energy*, u, is the sum of the potential and kinetic energy of all molecules in the system.

<u>Definition</u>: *Specific enthalpy*, h

$$h = u + p \cdot v \qquad [2.3]$$

or, in differential form

$$dh = du + p \cdot dv + v \cdot dp \qquad [2.4]$$

2.3 First Law of Thermodynamics

Energy cannot be created nor destroyed. Heat and work are equivalent.

2.4 Energy Balance for a Closed System

The first law of thermodynamics can, for a closed system, be formulated

$$q = \Delta u + \varepsilon_y = u_\beta - u_\alpha + \varepsilon_y \qquad [2.5]$$

or, in differential form

$$dq = du + d\varepsilon_y \qquad [2.6]$$

If the concept of friction work, $|b|$, is introduced, equations 2.5 and 2.6 can be written as

$$q + |b| = q_r = u_\beta - u_\alpha + \varepsilon_{yr} = u_\beta - u_\alpha + \int_\alpha^\beta p \cdot dv \qquad [2.7]$$

$$dq + |db| = dq_r = du + d\varepsilon_{yr} = du + p \cdot dv \qquad [2.8]$$

respectively. Introducing the definition of enthalpy, the first law of thermodynamics for a closed system can be written as

$$q_r = u_\beta - u_\alpha + \varepsilon_{yr} = u_\beta - u_\alpha + \int_\alpha^\beta p \cdot dv = h_\beta - h_\alpha - \int_\alpha^\beta v \cdot dp \qquad [2.9]$$

$$dq_r = du + d\varepsilon_{yr} = du + p \cdot dv = dh - v \cdot dp \qquad [2.10]$$

2.5 Energy Balance for an Open System

For *open systems at steady state*, the first law of thermodynamics can be written as

$$q = \varepsilon_t + h_\beta - h_\alpha + \frac{1}{2} \cdot (w_\beta^2 - w_\alpha^2) + g \cdot (z_\beta - z_\alpha) \qquad [2.11]$$

or, in differential form

$$dq = d\varepsilon_t + dh + w \cdot dw + g \cdot dz \qquad [2.12]$$

where ε_t : technical work per unit mass (work for open systems),

         w : velocity, $w = \dot{m}/(\rho \cdot A) = \dot{V}/A$,

         z : height above an arbitrarily chosen reference level,

         g : gravity acceleration constant ($= 9.81$ m/s²).

Note that no assumption of reversibility has been made, since $\varepsilon_t = \varepsilon_{tr} - |b|$

For reversible cases, the technical work can be found as

$$\varepsilon_{tr} = -\int_\alpha^\beta v \cdot dp + \frac{1}{2} \cdot (w_\alpha^2 - w_\beta^2) + g \cdot (z_\alpha - z_\beta) \qquad [2.13]$$

where the terms containing velocities and heights often can be neglected.

The first law of thermodynamics for *open systems under transient conditions*

$$Q = E_t + \int_{\tau_1}^{\tau_2} \left\{ \left[\dot{m} \cdot \left(h + \frac{w^2}{2} + g \cdot z \right) \right]_{outlet} - \left[\dot{m} \cdot \left(h + \frac{w^2}{2} + g \cdot z \right) \right]_{inlet} \right\} \cdot d\tau +$$

$$+ \left[m \cdot \left(u + \frac{w^2}{2} + g \cdot z \right) \right]_{\tau_2} - \left[m \cdot \left(u + \frac{w^2}{2} + g \cdot z \right) \right]_{\tau_1} \qquad [2.14]$$

where τ : time.

4

However, when evaluating equation 2.14, the *conservation of mass* has to be considered

$$\int_{\tau_1}^{\tau_2} \left(\dot{m}_{inlet} - \dot{m}_{outlet} \right) \cdot d\tau = m_{\tau_2} - m_{\tau_1} \qquad [2.15]$$

2.6 Energy Analysis of Cycles

Definition: A *cycle* is when a system has undergone a sequence of processes bringing the system back to its initial state. Two types of cycles exist. *Power cycles* which "produce" work and *Heat pump and refrigeration* cycles which require work.

Sign Convention
Heat: Positive if supplied to the system.
Work: Positive if rejected by the system.

Power cycles:

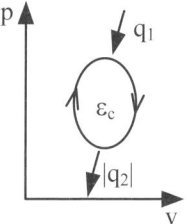

q_1 : Supplied heat,

$|q_2|$: Rejected heat,

ε_c : Rejected work.

$$\varepsilon_c = q_1 - |q_2| \qquad [2.16]$$

Thermal efficiency, η_t:

$$\eta_t = \frac{\varepsilon_c}{q_1} = \frac{q_1 - |q_2|}{q_1} = 1 - \frac{|q_2|}{q_1} \qquad [2.17]$$

Heat pump and refrigeration cycles:

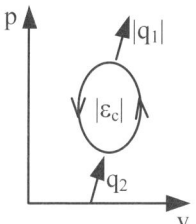

$|q_1|$: Rejected heat,

q_2 : Supplied heat,

$|\varepsilon_c|$: Supplied work.

$$|\varepsilon_c| = |q_1| - q_2 \qquad [2.18]$$

Coefficient of heating performance, COP$_1$:

$$COP_1 = \frac{|q_1|}{|\varepsilon_c|} \qquad [2.19]$$

Coefficient of cooling performance, COP$_2$:

$$COP_2 = \frac{q_2}{|\varepsilon_c|} \qquad [2.20]$$

Relation between COP$_1$ and COP$_2$

$$COP_1 = COP_2 + 1 \qquad [2.21]$$

3. Ideal Gas Model
3.1 Ideal Gas Law

$$p \cdot v = R \cdot T \qquad\qquad [3.1]$$

where p : absolute pressure (Pa),
v : specific volume (m³/kg),
R : individual gas constant (J/(kg·K)),
T : absolute temperature (K).

The individual gas constant, R, is defined as

$$R = \frac{R_M}{M} \qquad\qquad [3.2]$$

where R_M : universal gas constant (= 8314.3 J/(kmol·K)),
M : molecular weight or molar mass (kg/kmol).

3.2 Mixtures of Ideal Gases

<u>Definition</u>: *Mass fraction*, ξ_i

$$\xi_i = \frac{m_i}{m} = \frac{m_i}{m_1 + m_2 + ... + m_i + ... + m_n} = \frac{m_i}{\sum_i m_i} \qquad\qquad [3.3]$$

where m_i : mass of component i,
m : total mass of mixture.

<u>Definition</u>: *Mole fraction*, ψ_i

$$\psi_i = \frac{N_i}{N} = \frac{N_i}{N_1 + N_2 + ... + N_i + ... + N_n} = \frac{N_i}{\sum_i N_i} \qquad\qquad [3.4]$$

where N_i : number of moles of component i,
N : total number of moles.

<u>Definition</u>: *Mean molecular weight of mixture*, M_m

$$M_m = \frac{m}{N} = \sum_i (\psi_i \cdot M_i) = \frac{1}{\sum_i \frac{\xi_i}{M_i}} \qquad\qquad [3.5]$$

where M_i : molecular weight of component i.

Dalton law: the total pressure, p, equals the sum of the partial pressures, p_i, in the mixture

$$p = \sum_i p_i \qquad\qquad [3.6]$$

<u>Definition</u>: *Individual gas constant of a mixture, R_m*

$$R_m = \sum_i (\xi_i \cdot R_i) = \frac{R_M}{M_m} = \frac{8314.3}{M_m} \qquad\qquad [3.7]$$

where $\qquad R_i \qquad\qquad$: individual gas constant of component i.

$$R_i = \frac{R_M}{M_i} = \frac{8314.3}{M_i} \qquad\qquad [3.8]$$

<u>Definition</u>: *Volumetric fraction, v_i*

$$v_i = \frac{V_i}{V} = \frac{p_i}{p} \qquad\qquad [3.9]$$

where $\qquad V_i \qquad\qquad$: partial volume of component i,
$\qquad\qquad V \qquad\qquad$: total volume of mixture.

Relation between v_i, ψ_i, and ξ_i

$$v_i = \psi_i = \frac{M_m}{M_i} \cdot \xi_i = \frac{R_i}{R_m} \cdot \xi_i \qquad\qquad [3.10]$$

3.3 Specific Heat
<u>Definition</u>: *Specific heat at constant volume, c_v*

$$c_v = \left(\frac{\partial u}{\partial T}\right)_v \qquad\qquad [3.11]$$

<u>Definition</u>: *Specific heat at constant pressure, c_p*

$$c_p = \left(\frac{\partial h}{\partial T}\right)_p \qquad\qquad [3.12]$$

For ideal gases it can be shown mathematically that the internal energy, u, and the enthalpy, h, are functions of the temperature only. Hence

$$du = c_v \cdot dT \qquad\qquad [3.13]$$

$$dh = c_p \cdot dT \qquad\qquad [3.14]$$

Relation between c_p and c_v

$$c_p - c_v = R = \frac{R_M}{M} = \frac{8314.3}{M} \qquad [3.15]$$

Mean value of c_p and c_v with respect of temperature

$$\left[c_p\right]_{T_\alpha}^{T_\beta} = \frac{1}{T_\beta - T_\alpha} \cdot \int_{T_\alpha}^{T_\beta} c_p \cdot dT \qquad [3.16]$$

$$\left[c_v\right]_{T_\alpha}^{T_\beta} = \frac{1}{T_\beta - T_\alpha} \cdot \int_{T_\alpha}^{T_\beta} c_v \cdot dT \qquad [3.17]$$

Definition: *Specific heat ratio* (isentropic exponent), κ:

$$\kappa = \frac{c_p}{c_v} \qquad [3.18]$$

For ideal gases, the following relations can be applied

$$c_p = \frac{\kappa \cdot R}{\kappa - 1} = \frac{\kappa \cdot R_M}{M \cdot (\kappa - 1)} = \frac{\kappa \cdot 8314.3}{M \cdot (\kappa - 1)} \qquad [3.19]$$

$$c_v = \frac{R}{\kappa - 1} = \frac{R_M}{M \cdot (\kappa - 1)} = \frac{8314.3}{M \cdot (\kappa - 1)} \qquad [3.20]$$

For mixtures of ideal gases, the mean specific heats can be found by:

$$c_{v,m} = \sum_i \xi_i \cdot c_{v,i} \qquad [3.21]$$

$$c_{p,m} = \sum_i \xi_i \cdot c_{p,i} \qquad [3.22]$$

3.4 Kinetic Gas Theory

The basic equation of kinetic gas theory is

$$\frac{1}{2} \cdot m_m \cdot \overline{w^2} = \frac{3}{2} \cdot k_B \cdot T \qquad\qquad [3.23]$$

where
$\quad m_m \qquad$: mass of one molecule,
$\quad w \qquad\;\;$: velocity of the molecules,
$\quad k_B \qquad$: Boltzmann's constant ($= 1.38062 \cdot 10^{-23}$ J/K),
$\quad T \qquad\;\;$: absolute temperature of the gas.

The specific heats c_p and c_v can, according to the kinetic gas theory, be found as

$$c_p = (f + 2) \cdot \frac{R}{2} = (f + 2) \cdot \frac{R_M}{2 \cdot M} \qquad\qquad [3.24]$$

$$c_v = f \cdot \frac{R}{2} = f \cdot \frac{R_M}{2 \cdot M} \qquad\qquad [3.25]$$

where
$\quad f \qquad\qquad$: number of degrees of freedom for energy.

The specific heat ratio (isentropic exponent), κ, can then be written

$$\kappa = \frac{c_p}{c_v} = \frac{f + 2}{f} \qquad\qquad [3.26]$$

The number of degrees of freedom, f, is dependent on the constitution of the molecule and the temperature. For "normal" temperatures, the influence of the temperature is negligible and the number of degrees of freedom is given in the table below.

Number of atoms in the molecule	Number of degrees of freedom	κ
1	3	1.67
2	5	1.40
>2	6	1.33

For higher temperatures, the atoms in the molecules are vibrating and give rise to more degrees of freedom which in turn increases the specific heat. This effect is only present for gases with more than one atom per molecule (i.e. inert gases such as helium, neon, argon, krypton, xenon, and radon are unaffected). Furthermore, the effect is increasingly more important with increasing number of atoms per molecule.

4. Processes with Ideal Gases

4.1 Simple Processes

In this section we assume that changes in potential and kinetic energy are negligible. We also assume that the ideal gas has constant c_p and c_v (these can be calculated with equations 3.19 and 3.20 respectively), and that κ and M are known. We denote "α" as the state before the process, and "β" as the state after the process.

4.1.1 Isothermal Process $T_\alpha = T_\beta$

$$\frac{p_\beta}{p_\alpha} = \frac{v_\alpha}{v_\beta} \tag{4.1}$$

$$\varepsilon_{yr} = \varepsilon_{tr} = q_r = \frac{R_M \cdot T}{M} \cdot \ln\left(\frac{p_\alpha}{p_\beta}\right) \tag{4.2}$$

4.1.2 Isochoric Process $v_\alpha = v_\beta$

$$\frac{p_\beta}{p_\alpha} = \frac{T_\beta}{T_\alpha} \tag{4.3}$$

$$\varepsilon_{yr} = 0 \tag{4.4}$$

$$\varepsilon_{tr} = v \cdot (p_\alpha - p_\beta) = \frac{R_M}{M} \cdot (T_\alpha - T_\beta) \tag{4.5}$$

$$q_r = c_v \cdot (T_\beta - T_\alpha) \tag{4.6}$$

4.1.3 Isobaric Process $p_\alpha = p_\beta$

$$\frac{v_\beta}{v_\alpha} = \frac{T_\beta}{T_\alpha} \tag{4.7}$$

$$\varepsilon_{yr} = p \cdot (v_\beta - v_\alpha) = \frac{R_M}{M} \cdot (T_\beta - T_\alpha) \tag{4.8}$$

$$\varepsilon_{tr} = 0 \tag{4.9}$$

$$q_r = c_p \cdot (T_\beta - T_\alpha) \tag{4.10}$$

4.1.4 Isentropic Process $q_r = 0$, $p \cdot v^\kappa = $ constant

$$\frac{v_\beta}{v_\alpha} = \left(\frac{p_\alpha}{p_\beta}\right)^{\frac{1}{\kappa}} \qquad [4.11]$$

$$\frac{T_\beta}{T_\alpha} = \left(\frac{p_\beta}{p_\alpha}\right)^{\frac{\kappa-1}{\kappa}} \qquad [4.12]$$

$$\frac{T_\beta}{T_\alpha} = \left(\frac{v_\alpha}{v_\beta}\right)^{\kappa-1} \qquad [4.13]$$

$$\varepsilon_{yr} = c_v \cdot (T_\alpha - T_\beta) = c_v \cdot T_\alpha \cdot \left[1 - \left(\frac{p_\beta}{p_\alpha}\right)^{\frac{\kappa-1}{\kappa}}\right] \qquad [4.14]$$

$$\varepsilon_{tr} = \kappa \cdot \varepsilon_{yr} = c_p \cdot (T_\alpha - T_\beta) = c_p \cdot T_\alpha \cdot \left[1 - \left(\frac{p_\beta}{p_\alpha}\right)^{\frac{\kappa-1}{\kappa}}\right] \qquad [4.15]$$

4.1.5 Polytropic Process $p \cdot v^n = $ constant (n = polytropic exponent)

$$\frac{v_\beta}{v_\alpha} = \left(\frac{p_\alpha}{p_\beta}\right)^{\frac{1}{n}} \qquad [4.16]$$

$$\frac{T_\beta}{T_\alpha} = \left(\frac{p_\beta}{p_\alpha}\right)^{\frac{n-1}{n}} \qquad [4.17]$$

$$\frac{T_\beta}{T_\alpha} = \left(\frac{v_\alpha}{v_\beta}\right)^{n-1} \qquad [4.18]$$

$$\varepsilon_{yr} = \frac{R_M}{M \cdot (n-1)} \cdot (T_\alpha - T_\beta) = \frac{R_M \cdot T_\alpha}{M \cdot (n-1)} \cdot \left[1 - \left(\frac{p_\beta}{p_\alpha}\right)^{\frac{n-1}{n}}\right] \qquad [4.19]$$

$$\varepsilon_{tr} = n \cdot \varepsilon_{yr} = \frac{n \cdot R_M}{M \cdot (n-1)} \cdot (T_\alpha - T_\beta) = \frac{n \cdot R_M \cdot T_\alpha}{M \cdot (n-1)} \cdot \left[1 - \left(\frac{p_\beta}{p_\alpha}\right)^{\frac{n-1}{n}}\right] \qquad [4.20]$$

$$q_r = c_v \cdot \frac{(n-\kappa)}{(n-1)} \cdot (T_\beta - T_\alpha) \qquad [4.21]$$

4.2 Carnot Cycle

Reversible cycle between two constant temperatures.

Power cycle: (clockwise)

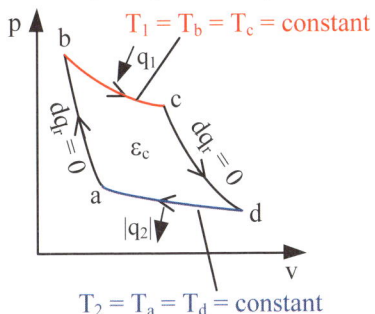

a→b : Isentropic compression,
b→c : Isothermal expansion,
c→d : Isentropic expansion,
d→a : Isothermal compression.

Thermal efficiency, η_{tC} :

$$\eta_{tC} = 1 - \frac{T_2}{T_1} = \eta_{t,Max} \qquad [4.22]$$

Heat pump and refrigeration cycle: (counterclockwise)

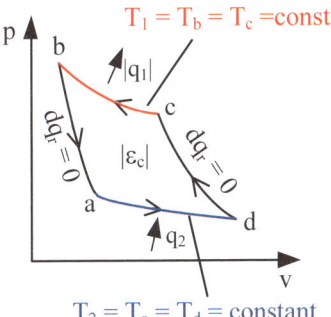

a→d : Isothermal expansion,
d→c : Isentropic compression,
c→b : Isothermal compression,
b→a : Isentropic expansion.

Coefficient of heating performance, COP_{1C}

$$COP_{1C} = \frac{T_1}{T_1 - T_2} = COP_{1,Max} \qquad [4.23]$$

Coefficient of cooling performance, COP_{2C}

$$COP_{2C} = \frac{T_2}{T_1 - T_2} = COP_{2,Max} \qquad [4.24]$$

Relation between COP_{1C} and COP_{2C}:

$$COP_{1C} = COP_{2C} + 1 \qquad [4.25]$$

4.3 Cycle Definitions

The compression ratio of a cycle, υ, is defined:

$$\upsilon = \frac{V_{largest}}{V_{smallest}} = \left\{ \begin{array}{l} v = V/m \\ m = constant \end{array} \right\} = \frac{v_{largest}}{v_{smallest}} \qquad [4.26]$$

where $\quad V_{largest}\quad$: largest volume in the cycle,
$\quad\quad\quad\quad V_{smallest}\quad$: smallest volume in the cycle,
$\quad\quad\quad\quad m\quad$: mass of circulating working media in the cycle,
$\quad\quad\quad\quad v_{largest}\quad$: largest specific volume in the cycle,
$\quad\quad\quad\quad v_{smallest}\quad$: smallest specific volume in the cycle.

The swept volume of a cycle, V_s, is defined:

$$V_s = V_{largest} - V_{smallest} \qquad [4.27]$$

The cycle work per unit mass, ε_c, is defined:

$$\varepsilon_c = q_1 - |q_2| = \oint \varepsilon_y = \oint \varepsilon_t = \sum_{cycle} \varepsilon_{y,process} = \sum_{cycle} \varepsilon_{t,process} \qquad [4.28]$$

Note that all heat flows supplied to and rejected from the cycle needs to be included in order for the first formulation of ε_c to be valid. The other formulations are always valid.

The cycle work, E_c, can be found by using equation 1.5:

$$E_c = m \cdot \varepsilon_c \qquad [4.29]$$

The net rate of cycle work, $\dot{E}_c$, can be calculated as:

$$\dot{E}_c = m \cdot \varepsilon_c \cdot \frac{n}{60} = E_c \cdot \frac{n}{60} \qquad [4.30]$$

where $\quad n \quad$: cycle (or rotational) speed per minute. I.e. the number of times the cycle completes per minute, [rpm].

The net rate of cycle work per unit swept volume, $\dfrac{\dot{E}_c}{V_s}$, can be calculated as:

$$\frac{\dot{E}_c}{V_s} = \frac{m \cdot \varepsilon_c}{\left(V_{largest} - V_{smallest} \right)} \cdot \frac{n}{60} = \frac{\varepsilon_c}{\left(v_{largest} - v_{smallest} \right)} \cdot \frac{n}{60} \qquad [4.31]$$

5. Second Law of Thermodynamics, Entropy and Exergy

5.1 Second Law of Thermodynamics

Heat cannot be transferred from a cold to a hot body spontaneously.

5.2 Entropy

The entropy describes the disorder in a system. As a system undergoes an irreversible process, the total entropy of the system changes. In general, the *entropy of a system increases if work is exchanged with the surroundings and/or heat is supplied to the system*. The only way to decrease the entropy of a system is by rejecting heat from it. It should however be noted that the entropy in the surroundings of the system is thereby increased. Further, the increase in entropy in the surroundings is larger than the decrease of entropy in the system.

Definition: *Entropy*, s

$$ds = \frac{dq_r}{T} \qquad [5.1]$$

Combining equation 5.1 with the first law of thermodynamics (equation 2.10)

$$ds = \frac{du + p \cdot dv}{T} = \frac{dh - v \cdot dp}{T} \qquad [5.2]$$

Assuming ideal gas behavior with constant c_p and c_v, equation 5.2 can be integrated

$$s_\beta - s_\alpha = c_v \cdot \ln\left(\frac{T_\beta}{T_\alpha}\right) + R \cdot \ln\left(\frac{v_\beta}{v_\alpha}\right)$$

$$s_\beta - s_\alpha = c_p \cdot \ln\left(\frac{T_\beta}{T_\alpha}\right) - R \cdot \ln\left(\frac{p_\beta}{p_\alpha}\right) \qquad [5.3]$$

$$s_\beta - s_\alpha = c_v \cdot \ln\left(\frac{p_\beta}{p_\alpha}\right) + c_p \cdot \ln\left(\frac{v_\beta}{v_\alpha}\right)$$

Equations 3.14 and 5.3 can be used to draw an s–T-diagram for an *ideal gas* with lines representing constant pressure, p, enthalpy, h, and specific volume, v.

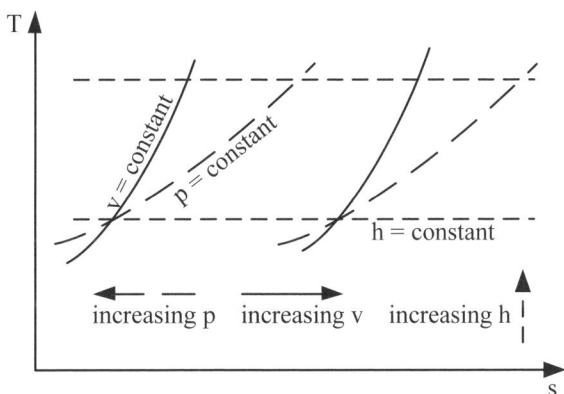

5.3 Carnot Cycle in an s–T-diagram

Power cycle | *Heat pump and refrigeration cycle*

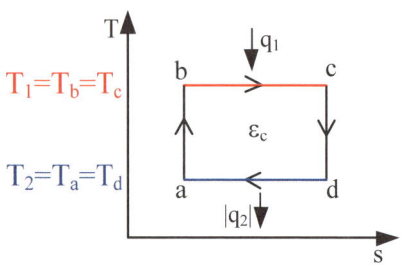

$$q_1 = \int_b^c T_1 \cdot ds = T_1 \cdot \Delta s \qquad [5.4]$$

$$|q_1| = \int_b^c T_1 \cdot ds = T_1 \cdot \Delta s \qquad [5.8]$$

$$|q_2| = \int_a^d T_2 \cdot ds = T_2 \cdot \Delta s \qquad [5.5]$$

$$q_2 = \int_a^d T_2 \cdot ds = T_2 \cdot \Delta s \qquad [5.9]$$

$$\varepsilon_c = q_1 - |q_2| \qquad [5.6]$$

$$|\varepsilon_c| = |q_1| - q_2 \qquad [5.10]$$

$$\eta_{tC} = \frac{\varepsilon_c}{q_1} = \frac{q_1 - |q_2|}{q_1} = \frac{T_1 - T_2}{T_1} =$$

$$COP_{2C} = \frac{q_2}{|\varepsilon_C|} = \frac{q_2}{|q_1| - q_2} = \frac{T_2}{T_1 - T_2} \qquad [5.11]$$

$$= 1 - \frac{T_2}{T_1} \qquad [5.7]$$

$$COP_{1C} = \frac{|q_1|}{|\varepsilon_C|} = \frac{|q_1|}{|q_1| - q_2} = \frac{T_1}{T_1 - T_2} \qquad [5.12]$$

5.4 Exergy (Availability)

Definition: *Exergy*, e, is the maximum amount of work available from a system interacting with the surroundings (index o). The definition of exergy can be written

$$e = q \cdot \frac{T - T_o}{T} \qquad [5.13]$$

Exergy of flow (flow availability) can be expressed

$$e = h - h_o + \frac{w^2}{2} + g \cdot z - T_o(s - s_o) \qquad [5.14]$$

or, if the kinetic and potential energies are neglected

$$e = h - h_o - T_o(s - s_o) \qquad [5.15]$$

For closed systems, the exergy, e_u, can be calculated using

$$e_u = u - u_o + p_o \cdot (v - v_o) - T_o(s - s_o) \qquad [5.16]$$

16

6. Gas Power Cycles

6.1 Otto Cycle

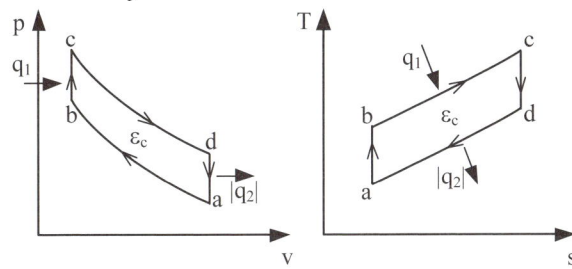

Thermal efficiency, η_t

$$\eta_t = 1 - \frac{(T_d - T_a)}{(T_c - T_b)} \qquad [6.1]$$

Compression Ratio, $\upsilon = v_a/v_b = V_a/V_b$

6.2 Diesel Cycle

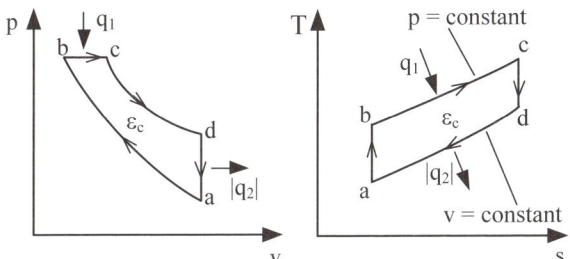

Thermal efficiency, η_t

$$\eta_t = 1 - \frac{1}{\kappa} \cdot \frac{(T_d - T_a)}{(T_c - T_b)} \qquad [6.2]$$

Compression Ratio, $\upsilon = v_a/v_b = V_a/V_b$

6.3 Joule/Brayton Cycle

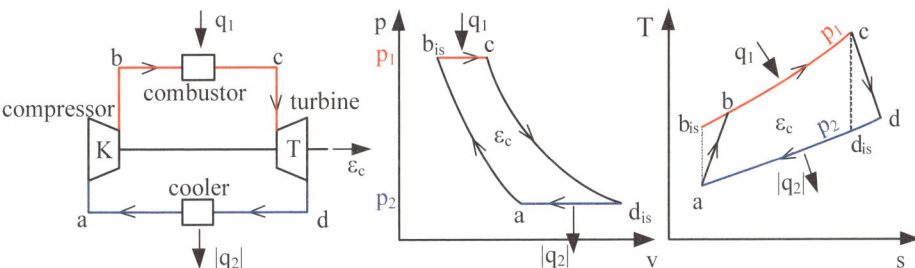

Thermal efficiency, η_t

$$\eta_t = \left(1 - \frac{1}{\tau}\right) \cdot \frac{\eta_T \cdot \dfrac{T_c}{T_a} - \dfrac{\tau}{\eta_K}}{\dfrac{T_c}{T_a} - 1 - \dfrac{\tau - 1}{\eta_K}} \qquad [6.3]$$

where $\qquad \tau = \left(\dfrac{p_1}{p_2}\right)^{\frac{\kappa - 1}{\kappa}}$

Definition: *Isentropic turbine efficiency*, η_T:

$$\eta_T = \frac{\varepsilon_T}{\varepsilon_{T,is}} = \frac{h_\alpha - h_\beta}{h_\alpha - h_{\beta is}} = \left\{\begin{matrix}\text{ideal gas} \\ \text{constant } c_p\end{matrix}\right\} = \frac{T_\alpha - T_\beta}{T_\alpha - T_{\beta is}} \qquad [6.4]$$

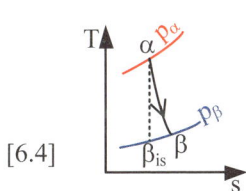

Definition: *Isentropic compressor efficiency*, η_K:

$$\eta_K = \frac{|\varepsilon_{K,is}|}{|\varepsilon_K|} = \frac{h_{\beta is} - h_\alpha}{h_\beta - h_\alpha} = \left\{ \begin{array}{l} \text{ideal gas} \\ \text{constant } c_p \end{array} \right\} = \frac{T_{\beta is} - T_\alpha}{T_\beta - T_\alpha} \qquad [6.5]$$

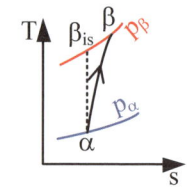

6.4 Ericsson Cycle

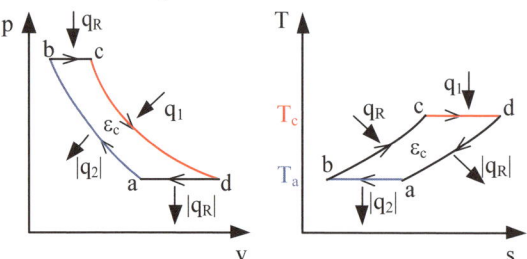

For an *ideal Ericsson cycle*, the thermal efficiency is

$$\eta_t = 1 - \frac{T_a}{T_c} \qquad [6.6]$$

which is equal to that of a Carnot cycle.

However, if the efficiency of the regenerator, η_R, is taken into account, the thermal efficiency can be expressed as

$$\eta_t = \frac{1}{1 + (1 - \eta_R) \cdot \dfrac{\kappa}{\kappa - 1} \cdot \left(1 - \dfrac{T_a}{T_c}\right) \cdot \dfrac{1}{\ln(p_1/p_2)}} \cdot \left(1 - \frac{T_a}{T_c}\right) \qquad [6.7]$$

6.5 Stirling Cycle

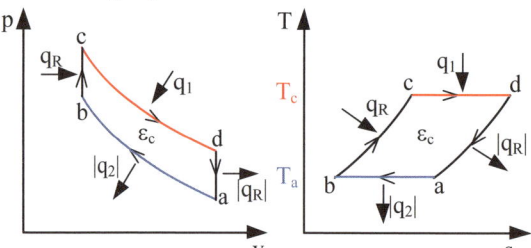

For an *ideal Stirling cycle*, the thermal efficiency is

$$\eta_t = 1 - \frac{T_a}{T_c} \qquad [6.8]$$

which is equal to that of a Carnot cycle.

However, if the efficiency of the regenerator, η_R, is taken into account, the thermal efficiency can be expressed as

$$\eta_t = \frac{1}{1 + \dfrac{(1 - \eta_R)}{\kappa - 1} \cdot \left(1 - \dfrac{T_a}{T_c}\right) \cdot \dfrac{1}{\ln(v_2/v_1)}} \cdot \left(1 - \frac{T_a}{T_c}\right) \qquad [6.9]$$

7. Thermodynamic Relations for Simple Compressible Substances
7.1 p–v–T-Surface

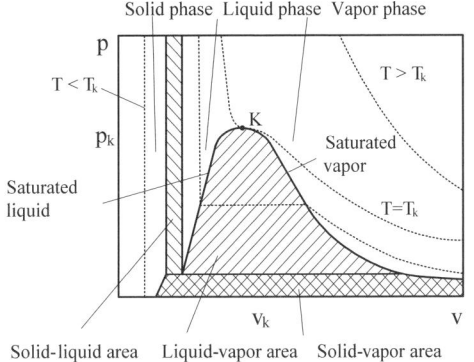
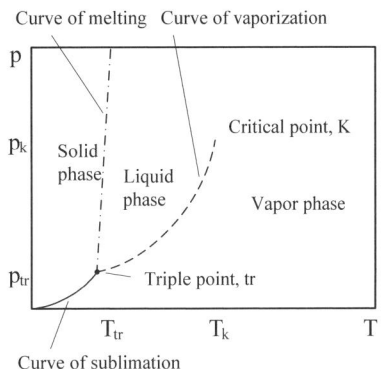

---- Isotherm, T=constant

The p-v-T-surface projected on the v, p- and T, p-plane respectively.

7.2 Two-phase Areas
<u>Definition</u>: *Vapor quality*, x:

$$x = \frac{m''}{m' + m''} = \frac{v_x - v'}{v'' - v'} = \frac{h_x - h'}{h'' - h'} = \frac{s_x - s'}{s'' - s'} \qquad [7.1]$$

where m'' : mass of saturated vapor,

            m' : mass of saturated liquid.

Specific volume of a liquid-vapor mixture with vapor quality, x

$$v_x = x \cdot v'' + (1 - x) \cdot v' \qquad [7.2]$$

where v'' : specific volume of saturated vapor,

            v' : specific volume of saturated liquid.

Enthalpy of a liquid-vapor mixture with vapor quality, x

$$h_x = x \cdot h'' + (1 - x) \cdot h' \qquad [7.3]$$

where h'' : enthalpy of saturated vapor,

            h' : enthalpy of saturated liquid.

Entropy of a liquid-vapor mixture with vapor quality, x

$$s_x = x \cdot s'' + (1 - x) \cdot s' \qquad [7.4]$$

where s'' : entropy of saturated vapor,

            s' : entropy of saturated liquid.

Definition: *Latent heat of vaporization*, r

$$r = h'' - h'$$ [7.5]

where   h''   : enthalpy of saturated vapor,
               h'   : enthalpy of saturated liquid.

Trouton's rule (approximates the latent heat of vaporization at a pressure of 1 atm, r_{1atm})

$$r_{1atm} = \frac{C_T \cdot T_{1atm}}{M}$$ [7.6]

where C_T : constant ($\approx$ 88 kJ/(kmol·K)),
               T_{1atm} : saturation temperature (K) at a pressure of 1 atm,
               M   : molecular weight (kg/kmol).

Clapeyron's equation (calculates the latent heat of vaporization, r)

$$r = (v'' - v') \cdot T \cdot \frac{dp}{dT}$$ [7.7]

where   v''   : specific volume of saturated vapor,
               v'   : specific volume of saturated liquid,
               T   : saturation temperature [K],
               $\dfrac{dp}{dT}$: slope of the vaporization curve (at temperature T) [Pa/K].

7.3 Generalized Compressibility Chart
Every medium can be described by the equation

$$p \cdot v = z_r \cdot R \cdot T$$ [7.8]

where z_r : compressibility factor.

z_r can be obtained from charts, $z_r = f$ (T, P). Each medium has its own chart. However, if the concept of reduced properties is introduced, all the charts, $z_r = f$ (T, P), for different media, can be generalized to *one single chart, $z_r = f$ (T_r, P_r)*, the **generalized compressibility chart**, see pages 51–53.

$$T_r = \frac{T}{T_k}$$ reduced temperature   [7.9]

$$P_r = \frac{p}{p_k}$$ reduced pressure   [7.10]

$$v_{ri} = \frac{v \cdot p_k}{R \cdot T_k}$$ pseudoreduced specific volume   [7.11]

where T_k and p_k are the critical temperature and pressure of the medium respectively. Data for T_k and p_k for a selection of gases can be found on page 43.

8. Vapor Power Systems
8.1 Rankine Process

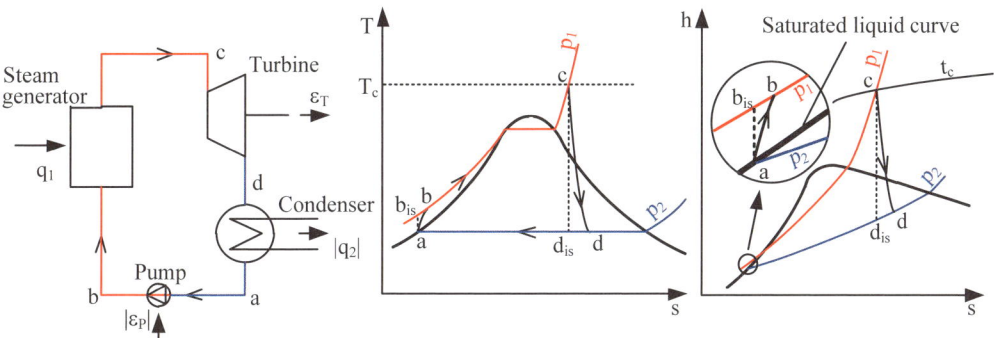

For the Rankine process, the following relations can be derived

$$\eta_T = \frac{h_c - h_d}{h_c - h_{dis}}$$ isentropic turbine efficiency [8.1]

$$\eta_P = \frac{h_{bis} - h_a}{h_b - h_a}$$ isentropic pump efficiency [8.2]

$$|\varepsilon_P| = h_b - h_a = \frac{v_a \cdot (p_b - p_a)}{\eta_P}$$ required pump work [8.3]

$$q_1 = h_c - h_b$$ required heat addition [8.4]

$$h_b = h_a + \frac{v_a \cdot (p_b - p_a)}{\eta_P}$$ enthalpy of state b [8.5]

$$\varepsilon_T = h_c - h_d = \eta_T \cdot (h_c - h_{dis})$$ turbine work [8.6]

$$|q_2| = h_d - h_a$$ rejected heat [8.7]

$$h_d = h_c - \eta_T \cdot (h_c - h_{dis})$$ enthalpy of state d [8.8]

$$\eta_t = \frac{\varepsilon_c}{q_1} = \frac{\varepsilon_T - |\varepsilon_P|}{q_1} = 1 - \frac{|q_2|}{q_1} = 1 - \frac{h_d - h_a}{h_c - h_b}$$ thermal efficiency [8.9]

8.2 Feedwater Heating

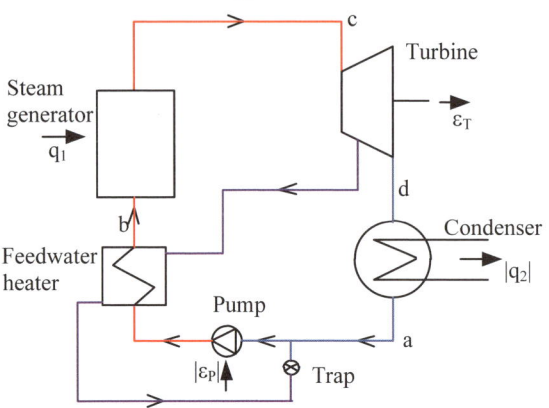

Media is extracted from the turbine to pre-heat the feedwater. Consequently, in the steam generator, the heat will be supplied to the system at a higher mean temperature resulting in *higher thermal efficiency.*

$$\eta_t = 1 - \frac{T_2}{T_{1,mean}} \qquad [8.10]$$

8.3 Reheating

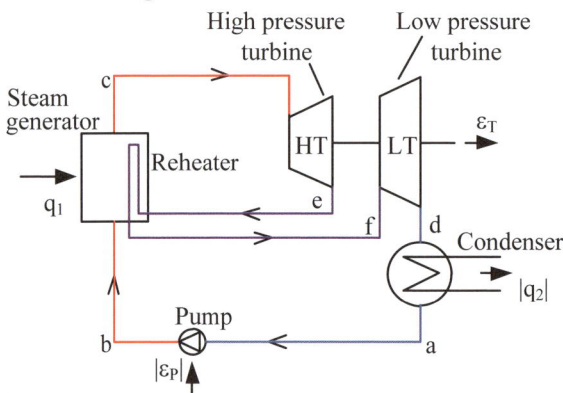

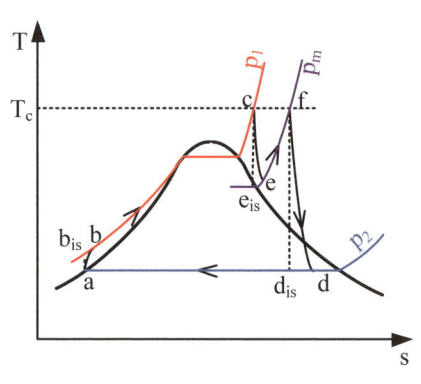

Although reheating gives a slight increase in thermal efficiency, it is used primarily to avoid *turbine blade erosion* caused by droplets condensing on the turbine blades. Turbine blade erosion will occur if the vapor quality is lower than 90 %. To avoid this, the expansion in the turbine is performed in two stages with reheating in between. By doing in this manner, process data can be chosen in such a way that the vapor quality at the turbine outlet (i.e. point d) is at least 90 %.

Since the expansion in the turbine is executed in two stages, an isentropic efficiency for each stage can be defined.

$$\eta_{T,HT} = \frac{h_c - h_e}{h_c - h_{eis}} \qquad \text{isentropic efficiency, high pressure turbine stage} \qquad [8.11]$$

$$\eta_{T,LT} = \frac{h_f - h_d}{h_f - h_{dis}} \qquad \text{isentropic efficiency, low pressure turbine stage} \qquad [8.12]$$

9. Refrigeration and Heat Pump Systems
9.1 Basic Vapor Compression Cycle

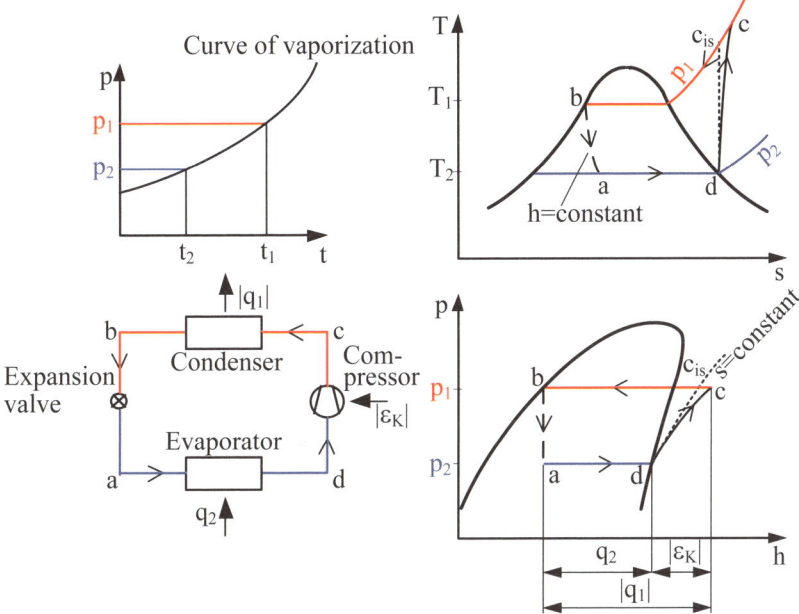

The vapor compression cycle (with isentropic compressor efficiency, η_K) without superheating and subcooling.

a-d: $\quad\quad q_2 = h_d - h_a$ [9.1]

d-c: $\quad\quad |\varepsilon_K| = h_c - h_d = \dfrac{h_{cis} - h_d}{\eta_K}$ [9.2]

c-b: $\quad\quad |q_1| = h_c - h_b$ [9.3]

b-a: $\quad\quad h_b = h_a$ [9.4]

Coefficient of cooling performance, COP_2

$$COP_2 = \frac{q_2}{|\varepsilon_K|} = \frac{h_d - h_a}{h_c - h_d} = \eta_K \cdot \frac{h_d - h_a}{h_{cis} - h_d} = \eta_K \cdot COP_{2d}$$ [9.5]

Coefficient of heating performance, COP_1

$$COP_1 = \frac{|q_1|}{|\varepsilon_K|} = \frac{h_c - h_b}{h_c - h_d} = \eta_K \cdot \frac{h_c - h_b}{h_{cis} - h_d} = COP_2 + 1$$ [9.6]

Carnot efficiency of refrigerant, η_{Cd}:

$$\eta_{Cd} = \frac{COP_{2d}}{COP_{2C}}$$ [9.7]

where $\quad\quad COP_{2C} = \dfrac{T_2}{T_1 - T_2}$, and $COP_{2d} = \dfrac{h_d - h_a}{h_{cis} - h_d}$.

9.2 Vapor Compression Cycle with Superheating and Subcooling

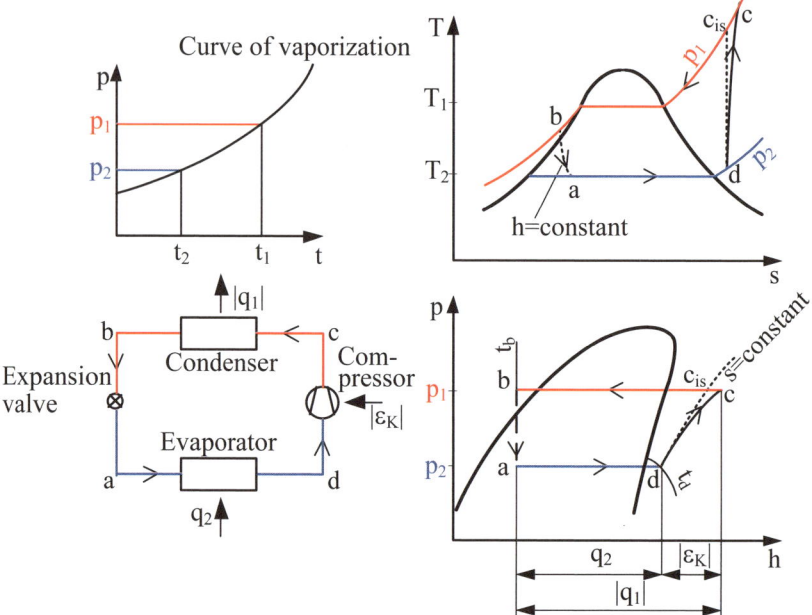

The subcooling, Δt_{sc}, is defined

$$\Delta t_{sc} = t_1 - t_b \qquad\qquad [9.8]$$

The superheating, Δt_{sh}, is defined

$$\Delta t_{sh} = t_d - t_2 \qquad\qquad [9.9]$$

where t_1 : condensation temperature (°C),
              t_2 : evaporation temperature (°C).

10. Fluid Mechanics

Conservation of mass can be written

$$\dot{m} = w_\alpha \cdot A_\alpha \cdot \rho_\alpha = w_\beta \cdot A_\beta \cdot \rho_\beta = w \cdot A \cdot \rho = \text{constant} \qquad [10.1]$$

10.1 Incompressible Fluid Mechanics

In incompressible fluid mechanics the fluid is assumed to have constant density

$$\rho = \frac{1}{v} = \text{constant} \qquad [10.2]$$

hence, equation 10.1 transforms to

$$\dot{V} = w_\alpha \cdot A_\alpha = w_\beta \cdot A_\beta = w \cdot A = \text{constant} \qquad [10.3]$$

where   A   : cross-sectional area.

Bernoulli's equation with losses

$$\frac{p_\alpha}{\rho} + \frac{w_\alpha^2}{2} + g \cdot z_\alpha = \frac{p_\beta}{\rho} + \frac{w_\beta^2}{2} + g \cdot z_\beta + \frac{\Delta p_f}{\rho} \qquad [10.4]$$

$$\Delta p_f = \Delta p_{fF} + \Delta p_{fS} \qquad [10.5]$$

where Δp_{fF} : frictional pressure drop,
                 Δp_{fS} : secondary pressure drop.

For flow in pipes

$$\Delta p_{fF} = f_1 \cdot \rho \cdot w^2 \cdot \frac{L}{d} \qquad [10.6]$$

$$\Delta p_{fS} = \zeta \cdot \frac{\rho \cdot w^2}{2} \qquad [10.7]$$

where f_1 : friction factor, see section 10.2,
                 L   : pipe length,
                 d   : pipe diameter,
                 ζ : loss coefficient, see section 10.3.

In a system consisting of multiple parts with different pipe diameter, the total pressure drop of the system is given by (the velocity in the different parts is obtained by equation 10.3)

$$\Delta p_f = \sum_m \left(f_1 \cdot \rho \cdot w^2 \cdot \frac{L}{d} \right)_m + \sum_n \left(\zeta \cdot \frac{\rho \cdot w^2}{2} \right)_n \qquad [10.8]$$

Two types of flow regimes are often distinguished, *laminar* and *turbulent*. To determine the flow regime, the Reynolds number, Re, is calculated.

Definition: *Reynolds number*, Re

$$Re = \frac{w \cdot d}{v} = \frac{w \cdot d \cdot \rho}{\mu}$$

[10.9]

where v : kinematic viscosity,
               μ : dynamic viscosity.

For non-circular tubes, the hydraulic diameter, d_h, is used:

$$d_h = \frac{4 \cdot A}{P}$$

[10.10]

where A : cross-sectional area of the tube,
               P : circumference of the tube.

10.2 Friction Factors

Laminar flow in tubes, Re < 2 300

$$f_l = \frac{C}{Re}$$

[10.11]

For *circular tubes*, C = 32. For other tube geometries, see table on page 65.

Turbulent flow (in circular tubes), Re > 2300
For turbulent flow in rough tubes, f_l can be found in the Moody chart (see page 61). For smooth tubes f_l can be found using one of the many correlations for f_l that exist. Below, a few of those are presented.

Blasius´ relation, 3 000 < Re < 100 000

$$f_l = 0.158 \cdot Re^{-0.25}$$

[10.12]

In the interval 5 000 < Re < 200 000, the following relation is often recommended

$$f_l = 0.092 \cdot Re^{-0.2}$$

[10.13]

According to Nikuradse, the following equation applies for $10^5 < Re < 10^7$

$$f_l = 0.0016 + 0.111 \cdot Re^{-0.237}$$

[10.14]

10.3 Loss Coefficients

Secondary losses occur in many different situations. Sudden changes of the pipe diameter and at passage through valves and bends, for instance. Although these losses are named "secondary losses" they can be quite substantial and has to be taken into account. A table of loss coefficients for common secondary losses can be found on pages 62–63.

10.4 Compressible Fluid Mechanics

Conservation of mass

$$\dot{m} = w_\alpha \cdot A_\alpha \cdot \rho_\alpha = w_\beta \cdot A_\beta \cdot \rho_\beta = w \cdot A \cdot \rho = \text{constant} \qquad [10.15]$$

Definition: *Velocity of sound*, a

$$a = \sqrt{\left(\frac{dp}{d\rho}\right)_s} = \left\{\begin{array}{l} \text{ideal} \\ \text{gas} \end{array}\right\} = \sqrt{\kappa \cdot R \cdot T} = \sqrt{\kappa \cdot p \cdot v} \qquad [10.16]$$

Definition: *Mach number*, Ma

$$Ma = \frac{w}{a} \qquad [10.17]$$

10.5 Hugoniot Equation

Assumes an ideal gas, isentropic process and no work exchange

$$\frac{dA}{A} = \left(Ma^2 - 1\right) \cdot \frac{dw}{w} \qquad [10.18]$$

Consequences of the Hugoniot equation. If

$\dfrac{dA}{A}$	$\left(Ma^2 - 1\right)$	$\dfrac{dw}{w}$
negative	positive $\Rightarrow$	negative. Deceleration !
negative	negative $\Rightarrow$	positive. Acceleration !
positive	positive $\Rightarrow$	positive. Acceleration !
positive	negative $\Rightarrow$	negative. Deceleration !

Further, Ma = 1 only if dA = 0. Observe, however, that dA = 0 *not necessarily has to imply* that Ma = 1, see sections 10.7–9.

10.6 Velocity in a Nozzle

For an ideal gas undergoing an isentropic process, the velocity, w_x, can be found by

$$w_x = \sqrt{\frac{2 \cdot \kappa}{\kappa - 1} \cdot R \cdot T_0 \cdot \left(1 - \left(\frac{p_x}{p_0}\right)^{\frac{\kappa - 1}{\kappa}}\right)} \qquad [10.19]$$

where x denotes the lengthwise position in the nozzle.

10.7 Critical Pressure Ratio

$p*/p_0$ is the pressure ratio that accelerates an ideal gas from standstill to the velocity of sound.

$$\frac{p*}{p_0} = \left(\frac{2}{\kappa + 1}\right)^{\frac{\kappa}{\kappa - 1}}$$

[10.20]

If $\dfrac{p*}{p_0} \geq \dfrac{p_2}{p_0}$, then $Ma \geq 1$.

10.8 Simple Nozzles

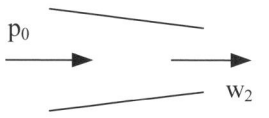

p_0 p_2 A simple nozzle consists only of a converging duct. Hence, according to Hugoniot´s equation, the highest velocity in a simple nozzle is the velocity of sound. If $p* \geq p_2$ then $Ma = 1$ at the outlet. If $p* < p_2$, then $Ma < 1$ at the outlet.

Simple nozzle

10.9 De Laval Nozzles

To achieve $Ma > 1$, a **De Laval nozzle** is required. The de Laval nozzle consists of a converging duct followed by a diverging duct.

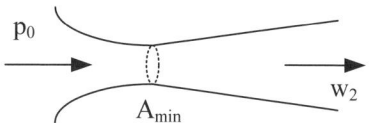

p_0 p_2 For a De Laval nozzle, designed in such a way that no shock losses occur, and $p* > p_2$, $Ma > 1$ at the outlet. Further, $Ma = 1$ at the position where $dA = 0$, i.e. at A_{min}.

de Laval nozzle

10.10 Media Flow of an Ideal Gas in a Nozzle

$$\dot{m} = A_x \cdot \psi_x \cdot \frac{p_0}{\sqrt{R \cdot T_0}}$$

[10.21]

$$\psi_x = \sqrt{\frac{2 \cdot \kappa}{\kappa - 1} \cdot \left(\frac{p_x}{p_0}\right)^{\frac{2}{\kappa}} \cdot \left(1 - \left(\frac{p_x}{p_0}\right)^{\frac{\kappa - 1}{\kappa}}\right)}$$

[10.22]

If $Ma = 1$ at the narrowest part of the duct (i.e. at A_{min}), then

$$\psi* = \sqrt{\kappa \cdot \left(\frac{2}{\kappa + 1}\right)^{\frac{\kappa + 1}{\kappa - 1}}}$$

[10.23]

10.11 Design of De Laval Nozzles for Ideal Gases

$$A_{min} = \frac{\dot{m} \cdot \sqrt{R \cdot T_0}}{\psi^* \cdot p_0}$$ [10.24]

$$\frac{A_x}{A_{min}} = \frac{\psi^*}{\psi_x} = f\left(\kappa, \frac{p_x}{p_0}\right)$$ [10.25]

10.12 Expansion Processes in Nozzles

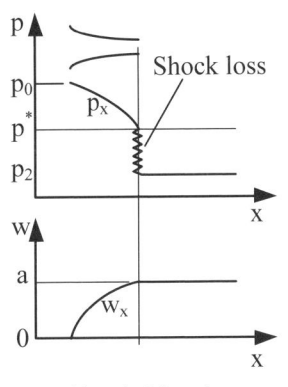

Simple Nozzle

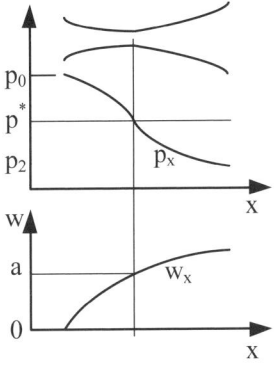

de Laval Nozzle

11. Heat Transfer

11.1 Conduction

(occurs primarily in solid or stagnant media)

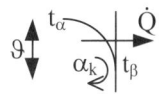

For steady state conduction heat transfer, the heat flow is given by

$$\dot{Q} = \frac{\lambda}{\delta} \cdot A \cdot \vartheta_v \qquad\qquad [11.1]$$

where
λ : thermal conductivity of the medium, i.e. a material property,
δ : thickness of the medium (in the direction of the heat flow),
A : heat transfer area (perpendicular to the direction of heat flow),
ϑ_v : temperature difference.

11.2 Convection

(occurs at the interaction between a flowing medium and a solid wall)

$$\dot{Q} = \alpha_k \cdot A \cdot \vartheta \qquad\qquad [11.2]$$

where
α_k : heat transfer coefficient,
A : heat transfer area (perpendicular to the direction of heat flow),
ϑ : temperature difference.

In order to determine the heat transfer coefficient for different cases, experimental investigations and/or theoretical work has been conducted. In these, the heat transfer coefficient is often expressed in a dimensionless form. The dimensionless number containing the heat transfer coefficient is called the Nusselt number, Nu.

$$Nu = \frac{\alpha_k \cdot d}{\lambda} \qquad\qquad [11.3]$$

where
d : characteristic length, for circular tubes d is the diameter,
λ : thermal conductivity of fluid.

As can be seen in equation 11.2 above, the heat transfer coefficient (α) is connected to the temperature difference (ϑ). Hence, the heat transfer coefficient (and as a consequence the Nusselt number) can be defined in different ways depending on the temperature difference used.

Two types of convection can be distinguished; *natural* and *forced* convection. In natural convection, the media flow is caused by differences in density which in turn is a result of the heat transfer itself. In forced convection, the flow of media is caused by some mechanical device such as a fan or a pump.

11.2.1 Natural Convection

In order to determine whether the flow is laminar or turbulent, the *Grashof number*, Gr, is introduced.

$$Gr = \frac{g \cdot \beta \cdot \Delta t \cdot H^3}{v^2} \qquad [11.4]$$

where
- g : gravity acceleration constant (= 9.81 m/s²),
- β : volume expansion coefficient of medium (for ideal gases, $\beta = 1/T$),
- Δt : temperature difference between flowing medium and solid,
- H : characteristic length (for vertical walls, H = wall height),
- v : kinematic viscosity of medium.

It can be shown that another dimensionless parameter, the *Prandtl number*, Pr, also influences the heat transfer.

$$Pr = \frac{\mu \cdot c_p}{\lambda} \qquad [11.5]$$

where
- μ : dynamic viscosity of medium,
- c_p : specific heat at constant pressure of medium,
- λ : thermal conductivity of medium.

Whether the flow is laminar or turbulent is determined by the following conditions
laminar: $10^4 < Gr \cdot Pr < 10^8$
turbulent: $10^8 < Gr \cdot Pr < 10^{12}$

For *vertical walls*, the heat transfer coefficient, α_k, can be calculated as
for $10^4 < Gr \cdot Pr < 10^8$

$$\alpha_k = \frac{\lambda}{H} \cdot Nu = \frac{\lambda}{H} \cdot 0.56 \cdot (Gr \cdot Pr)^{\frac{1}{4}} \qquad [11.6]$$

for $10^8 < Gr \cdot Pr < 10^{12}$

$$\alpha_k = \frac{\lambda}{H} \cdot Nu = \frac{\lambda}{H} \cdot 0.13 \cdot (Gr \cdot Pr)^{\frac{1}{3}} \qquad [11.7]$$

As can be seen, the product Gr·Pr is commonly used in natural convection. Actually, in some texts the product Gr·Pr is called the Rayleigh Number, Ra. The product Gr·Pr can be written:

$$Gr \cdot Pr = \frac{g \cdot \beta \cdot \Delta t \cdot H^3}{v^2} \cdot \frac{\mu \cdot c_p}{\lambda} = \frac{g \cdot \beta \cdot \mu \cdot c_p}{v^2 \cdot \lambda} \cdot \Delta t \cdot H^3 = C_{Ra} \cdot \Delta t \cdot H^3 \quad [11.8]$$

where C_{Ra} : temperature-dependent function, "Rayleigh function".

The "Rayleigh function", C_{Ra}, is temperature-dependent and should be evaluated at $t_{film} = (t_{amb} + t_{surface})/2$. Data for C_{Ra} for water and air can be found in tables on pages 48–49.

11.2.2 Forced Convection

In forced convection, the Reynolds number, Re, is used to determine whether the flow is laminar or turbulent

$$Re = \frac{w \cdot d}{\nu} = \frac{w \cdot d \cdot \rho}{\mu} \qquad [11.9]$$

For heat transfer in pipes, the following conditions apply

laminar: $Re < 2\,300$
turbulent: $2\,300 < Re$

For *laminar cases*, it has been shown that the heat transfer coefficient becomes constant after a certain distance from the inlet. This distance, the so-called *thermal entry length*, L_t, can be calculated as

$$L_t = 0.05 \cdot Re \cdot Pr \cdot d \qquad [11.10]$$

where   d   : pipe diameter.

For long pipes, $L_t/L < 0.05$ (or Gz < 1), the heat transfer coefficient can therefore be assumed constant. For pipes with *circular cross-section* and *constant surface temperature*, the heat transfer coefficient is given by

$$\alpha_{ln,Lam} = \frac{\lambda}{d} \cdot Nu_{ln} = 3.66 \cdot \frac{\lambda}{d} \qquad [11.11]$$

For other types of cross-sections and boundary conditions, the reader is referred to page 65.

For short pipes of different geometries, the heat transfer coefficient can be calculated using:

$$\alpha_{ln,Lam} = \frac{\lambda}{d} \cdot Nu_{ln} = \frac{\lambda}{d} \cdot \left(Nu_{Ref} + \frac{0.0298 \cdot Gz^{1.37}}{1 + 0.0438 \cdot Gz^{0.87}} \right) \qquad [11.12]$$

where Nu_{Ref}:   reference Nusselt number, dependent on geometry and boundary
                                condition (see page 65),
             Gz   Graetz number, $Gz = Re \cdot Pr \cdot d / L$.

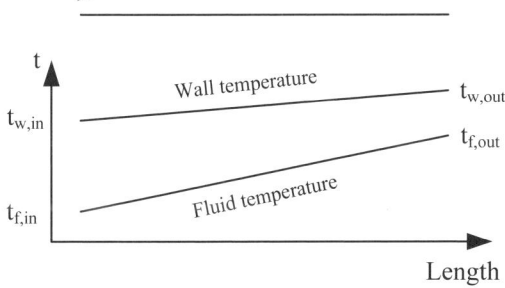

$$\dot{Q} = \alpha_{ln,Lam} \cdot A \cdot \vartheta_{ln}$$

$$\vartheta_{ln} = \frac{(t_{w,in} - t_{f,in}) - (t_{w,out} - t_{f,out})}{\ln\left(\dfrac{t_{w,in} - t_{f,in}}{t_{w,out} - t_{f,out}} \right)}$$

For *turbulent cases*, the heat transfer coefficient can be calculated by the "classic" correlation proposed by Dittus and Boelter

$$\alpha_{k,Turb} = \frac{\lambda}{d} \cdot Nu = \frac{\lambda}{d} \cdot 0.023 \cdot Re^{0.8} \cdot Pr^{0.4} \qquad [11.13]$$

11.3 Radiation

The emitted energy rate from a black body by electromagnetic radiation, Φ_{black}, is given by the Stefan-Boltzmann law

$$\Phi_{black} = \sigma \cdot A \cdot T^4 \qquad [11.14]$$

where σ : Stefan-Boltzmann constant ($= 5.67 \cdot 10^{-8}$ W/(m$^2 \cdot$K^4)).

Definition: A *black body* absorbs all incoming electromagnetic radiation.

For a non-black body (a grey body) the Stefan-Boltzmann law is corrected according to

$$\Phi_{grey} = \varepsilon \cdot \sigma \cdot A \cdot T^4 \qquad [11.15]$$

where ε : emissivity, $\varepsilon < 1$, see table on page 50.

Two bodies which "see" each other, have a net heat exchange (index $\alpha\beta$ denotes a net exchange from body α to body β)

$$\dot{Q}_{\alpha\beta} = \sigma \cdot F_{\alpha\beta} \cdot A_\alpha \cdot \left(T_\alpha^4 - T_\beta^4\right) \qquad [11.16]$$

where $F_{\alpha\beta}$: geometry factor.

For plane parallel plates ($A_\alpha = A_\beta$):

$$F_{\alpha\beta} = \frac{1}{\dfrac{1}{\varepsilon_\alpha} + \dfrac{1}{\varepsilon_\beta} - 1} \qquad [11.17]$$

For concentric bodies (with $A_\alpha \ll A_\beta$), $F_{\alpha\beta} = \varepsilon_\alpha$.

A radiation heat transfer coefficient, α_s, can be defined as

$$\alpha_s = \sigma \cdot F_{\alpha\beta} \cdot \left(\frac{T_\alpha^4 - T_\beta^4}{t_\alpha - t_\beta}\right) \qquad [11.18]$$

then, the heat exchange by radiation can be written as

$$\dot{Q}_s = \alpha_s \cdot A \cdot (t_\alpha - t_\beta) = \alpha_s \cdot A \cdot \vartheta \qquad [11.19]$$

11.4 Heat Exchangers

Three types of heat exchangers are distinguished:

- Recuperative,
- Regenerative (Ljungström preheater),
- Evaporative (Cooling towers).

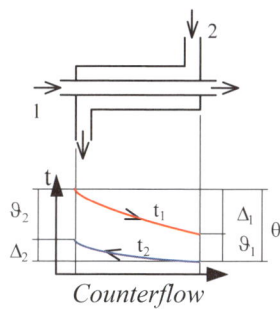
Counterflow

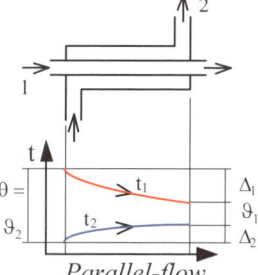
Parallel-flow

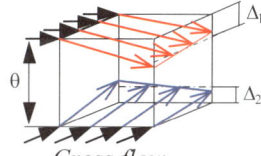
Cross flow

The *recuperative* heat exchangers can be divided into:

- Counterflow heat exchangers
- Parallel-flow heat exchangers
- Cross flow heat exchangers

For calculations, the following equations apply:

$$\dot{Q} = (\dot{m} \cdot c_p \cdot \Delta)_1 = (\dot{m} \cdot c_p \cdot \Delta)_2 \qquad [11.20]$$

$$\dot{Q} = k \cdot A \cdot \vartheta_m \qquad [11.21]$$

$$\vartheta_m = \frac{\vartheta_1 - \vartheta_2}{\ln(\vartheta_1/\vartheta_2)} \qquad [11.22]$$

The temperature efficiencies are defined:

$$\eta_1 = \frac{\Delta_1}{\theta} \qquad [11.23]$$

$$\eta_2 = \frac{\Delta_2}{\theta} \qquad [11.24]$$

Introduce:

$$\dot{W}_1 = (\dot{m} \cdot c_p)_1 \qquad [11.25]$$

$$\dot{W}_2 = (\dot{m} \cdot c_p)_2 \qquad [11.26]$$

hence

$$\eta_2 = \eta_1 \cdot \frac{\dot{W}_1}{\dot{W}_2} = Y \cdot \eta_1 \qquad [11.27]$$

For *counterflow heat exchangers* it can be shown that

$$\eta_1 = \frac{1 - e^{-X \cdot (1-Y)}}{1 - Y \cdot e^{-X \cdot (1-Y)}} \qquad \text{(For } Y = 1, \ \eta_1 = X/(X+1)) \qquad [11.28]$$

where

$$X = \frac{k \cdot A}{\dot{W}_1} \qquad [11.29]$$

$$Y = \frac{\dot{W}_1}{\dot{W}_2} \qquad\qquad [11.30]$$

For *parallel-flow heat exchangers* it can be shown that

$$\eta_1 = \frac{1 - e^{-X \cdot (1+Y)}}{1 + Y} \qquad\qquad [11.31]$$

To be able to use the diagrams on page 64, set $\dot{W}_1 < \dot{W}_2$.

For *cross flow heat exchangers* matters are a bit more complicated. For solving these kind of problems, the reader is referred to **Compact Heat Exchangers** by W. M. Kays and A. L. London, 1964.

11.5 Heat Transfer through Walls

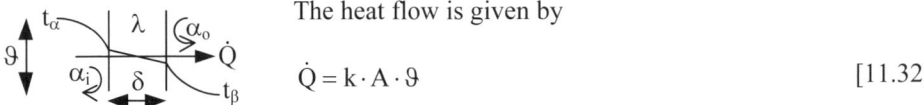

The heat flow is given by

$$\dot{Q} = k \cdot A \cdot \vartheta \qquad\qquad [11.32]$$

where   k   : overall heat transfer coefficient,
         ϑ : temperature difference.

The overall heat transfer coefficient for a plane wall consisting of multiple layers is calculated as

$$\frac{1}{k} = \frac{1}{\alpha_i} + \sum_n \left(\frac{\delta}{\lambda}\right)_n + \frac{1}{\alpha_o} \qquad\qquad [11.33]$$

where α_i : heat transfer coefficient on the inside (convection + radiation),
         α_o : heat transfer coefficient on the outside (convection + radiation).

For cylindrical and spherical walls consisting of multiple layers, the overall heat transfer coefficient can be found by (A_i and A_o are the inside and outside surface areas respectively)

$$\frac{1}{k \cdot A} = \frac{1}{\alpha_i \cdot A_i} + \sum_n \left(\frac{\delta}{\lambda \cdot A_m}\right)_n + \frac{1}{\alpha_o \cdot A_o} \qquad\qquad [11.34]$$

where, for *cylindrical walls* (r_i: inside radius and r_o: outside radius)

$$A_m = \frac{2 \cdot \pi \cdot L \cdot (r_o - r_i)}{\ln(r_o / r_i)} \qquad\qquad [11.35]$$

for *spherical walls* (r_i: inside radius and r_o: outside radius)

$$A_m = 4 \cdot \pi \cdot r_o \cdot r_i \qquad\qquad [11.36]$$

12. Psychrometrics
12.1 Introduction

<u>Definition</u>: *Relative humidity*, φ:

$$\varphi = \frac{p_v}{p_v''} \qquad\qquad\qquad [12.1]$$

where
$\quad p_v \qquad$: partial pressure of water vapor in air,
$\quad p_v'' \qquad$: saturation pressure of water at air temperature.

The saturation pressure of water, p_v'', can according to Granryd be found as

$$p_v'' = e^{\left(12.03 - \frac{4025}{t+235}\right)} \qquad \text{for } t \geq 0\ ^\circ\text{C} \qquad \text{(bar)} \qquad [12.2]$$

$$p_v'' = e^{\left(17.391 - \frac{6142.83}{t+273.15}\right)} \qquad \text{for } -40 < t \leq 0\ ^\circ\text{C} \qquad \text{(bar)} \qquad [12.3]$$

where
$\quad t \qquad$: temperature (°C).

<u>Definition</u>: *Water content*, x_w

$$x_w = \frac{m_{H_2O}}{m_{Air}} = \frac{m_v}{m_A} \qquad\qquad \text{(kg } H_2O/\text{kg dry air)} \ [12.4]$$

where
$\quad m_v \qquad$: mass of water vapor in air,
$\quad m_A \qquad$: mass of dry air.

By aid of the ideal gas law, equation 12.4 can be written as

$$x_w = \frac{M_v}{M_A} \cdot \frac{p_v}{p_{tot} - p_v} \approx 0.622 \cdot \frac{p_v}{p_{tot} - p_v} \qquad \text{(kg } H_2O/\text{kg dry air)} \ [12.5]$$

where
$\quad M_v \qquad$: molecular weight of water vapor ($= 18$ kg/kmol),
$\quad M_A \qquad$: molecular weight of air (≈ 29 kg/kmol),
$\quad p_{tot} \qquad$: total pressure of humid air.

Assuming ideal gas behavior, the enthalpy of humid air, h_{1+x}, can be found as

$$h_{1+x} = h_A + x_w \cdot h_v = t \cdot c_{p,A} + x_w \cdot \left(r + c_{p,v} \cdot t\right) \qquad \text{(kJ/kg dry air)} \quad [12.6]$$

where
$\quad h_A \qquad$: enthalpy of dry air,
$\quad h_v \qquad$: enthalpy of water vapor,
$\quad t \qquad$: temperature,
$\quad c_{p,A} \qquad$: specific heat of air (≈ 1 kJ/(kg·K)),
$\quad r \qquad$: latent heat of vaporization of water ($\approx 2\ 500$ kJ/kg),
$\quad c_{p,v} \qquad$: specific heat of water vapor (≈ 1.86 kJ/(kg·K)).

Hence, equation 12.6 can be written

$$h_{1+x} \approx t + x_w \cdot (2500 + 1.86 \cdot t) \qquad \text{(kJ/kg dry air) [12.7]}$$

Definition: *Dew point*, t_{dew}, is the temperature at which $p_v'' = p_v$.

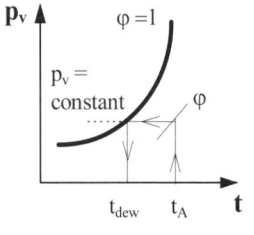
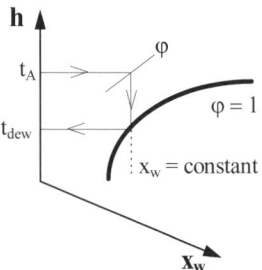

12.2 Steady State Mixing of Two Streams of Humid Air
Objective: find h_{bl}, $x_{w,bl}$, m_{bl}

<u>Analytically:</u>

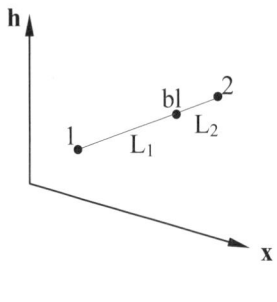

$m_1, h_1, x_{w,1}$

$m_{bl}, h_{bl}, x_{w,bl}$

$m_2, h_2, x_{w,2}$

Mixing of two streams of humid air

$$m_{bl} = m_1 + m_2 \qquad [12.8]$$

$$x_{w,bl} = \frac{(m \cdot x_w)_1 + (m \cdot x_w)_2}{m_{bl}} \qquad [12.9]$$

$$h_{bl} = \frac{(m \cdot h)_1 + (m \cdot h)_2}{m_{bl}} \qquad \text{(kJ/kg dry air)[12.10]}$$

<u>Graphically</u>

$$\frac{m_1}{m_2} = \frac{x_{w,bl} - x_{w,2}}{x_{w,1} - x_{w,bl}} = \frac{h_{bl} - h_2}{h_1 - h_{bl}} \qquad [12.11]$$

The "mixing point" lies on a straight line between points 1 and 2. Measure the distance $(L_1 + L_2)$ in the graph. Then

$$L_1 = \frac{m_2}{m_1 + m_2} \cdot (L_1 + L_2) \qquad [12.12]$$

and the "mixing point" , bl, can be found in the graph.

12.3 Mass Transfer Influence on Heat Transfer

Humid air
t_A, p_v

Moist surface, t'', p_v''

Height above surface

$p_v > p_v''$

$\dot{m}_d$

p_v'' p_v p

Height above surface

$p_v < p_v''$

$\dot{m}_d$

p_v p_v'' p

Since water either evaporates or condenses, heat must be either supplied to or rejected from the surface. The heat flow due to diffusion, $\dot{Q}_d$, is given by

$$\dot{Q}_d = \dot{m}_d \cdot r \qquad \text{for } t'' > 0\,°\text{C}$$
$$\dot{Q}_d = \dot{m}_d \cdot (r + \ell) \qquad \text{for } t'' \leq 0\,°\text{C} \qquad\qquad [12.13]$$

$$r = (h'' - h') \qquad (\approx 2500 \text{ kJ/kg for water}) \qquad\qquad [12.14]$$

where $\dot{m}_d$: diffusion mass flow rate of water,
        ℓ : latent heat of melting (≈ 335 kJ/kg for water).

The mass transfer can be described as

$$\dot{m}_d = \delta \cdot A \cdot (p_v'' - p_v) = \beta \cdot A \cdot (\rho_v'' - \rho_v) = \sigma \cdot A \cdot (x_w'' - x_w) \qquad [12.15]$$

where δ, β and σ : different mass transfer coefficients.

Assuming ideal gas behavior, the relation between the mass transfer coefficients can be written

$$\delta \cdot (p_v'' - p_v) = \beta \cdot \left(\frac{p_v''}{R_v \cdot T} - \frac{p_v}{R_v \cdot T} \right) \approx \sigma \cdot \frac{M_v}{M_A} \cdot \frac{(p_v'' - p_v)}{p_{A,m}} \qquad [12.16]$$

where $p_{A,m}$: mean partial pressure of dry air above the moist surface.

12.4 Lewis Relation

$$\sigma = \frac{\alpha_k}{c_{p,1+x}} \qquad\qquad [12.17]$$

where α_k : convective heat transfer coefficient,
        $c_{p,\,1+x}$: specific heat of humid air.

$$c_{p,1+x} = c_{p,A} + x_{w,m} \cdot c_{p,v} \qquad\qquad [12.18]$$

where: $x_{w,m}$: mean water content in the air and above the surface.

$$x_{w,m} = \frac{x_w'' + x_w}{2} = 0.622 \cdot \frac{\dfrac{p_v + p_v''}{2}}{p_{tot} - \dfrac{p_v + p_v''}{2}} \qquad [12.19]$$

12.5 Bäckström Relation

$$\frac{\alpha_d}{\alpha_k} = C_B \cdot \frac{p_v - p_v''}{t_A - t''} \qquad [12.20]$$

where $\quad \alpha_d \qquad$: diffusion heat transfer coefficient,

$$C_B \qquad : \text{constant} \begin{cases} = 1520\,(°C/bar)\,\text{if}\ \ t'' > 0°C \\ = 1750\,(°C/bar)\,\text{if}\ \ t'' \le 0°C \end{cases}$$

$$\dot{Q}_d = \alpha_d \cdot A \cdot (t_A - t'') \qquad [12.21]$$

12.6 Total Heat Transfer

For a system where heat is transferred by convection, diffusion and radiation, the total heat flow is found by

$$\dot{Q}_{tot} = \left[\alpha_k \cdot \left(1 + \frac{\alpha_d}{\alpha_k}\right) + \alpha_s \right] \cdot A \cdot (t_A - t'') \qquad [12.22]$$

where $\dfrac{\alpha_d}{\alpha_k}$ can be calculated by the Bäckström relation.

It should be noted that α_d / α_k in certain cases can assume negative values. This phenomenon is explained by the fact that the *heat transfer associated with the mass transfer is going in the opposite direction than that of the convective heat transfer.* At thermal equilibrium between a moist surface and the surrounding air (in a case where the influence of radiation can be neglected), this ratio will be equal to –1. The temperature of the moist surface will in this case be equal to the wet bulb temperature, $t_{wet\ bulb}$.

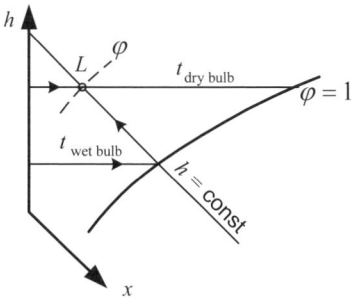

Tables and Diagrams

	Page
Molar Mass, and Critical-point Properties for a Selection of Gases	43
Approximate Molar Mass for a selection of substances	43
Thermodynamic Properties of Saturated Water	44
Thermodynamic Properties of Saturated Nitrogen	45
Thermodynamic Properties of Saturated Carbon Dioxide	46
Thermodynamic Properties of Saturated Ammonia	47
Thermodynamic Properties of Saturated R22	47
Thermodynamic Properties for Solid, Liquid, and Gaseous Water	48
Thermodynamic Properties for Air	49
Emissivity (and absorptivity)	50
Generalized Compressibility Diagram	51
Temperature – Entropy Diagram for Water	54
Temperature – Entropy Diagram for Nitrogen	55
Temperature – Entropy Diagram for Carbon Dioxide	56
Enthalpy – Entropy Diagram for Water	57
Enthalpy – Entropy Diagram for Air	58
Pressure – Enthalpy Diagram for Ammonia	59
Pressure – Enthalpy Diagram for R22	60
Moody Chart	61
Pressure Loss Coefficients for Inlets, Valves, Bends, and Knees	62
Countercurrent/Parallel-flow Heat Exchanger Diagram	64
Nusselt Number and Friction Coefficient for Laminar Flow in Long Ducts	65
Partial Pressure – Temperature Diagram for Humid Air	66
Enthalpy – Water Content Diagram for Humid Air	67
Unit Conversion and Prefixes	68
Mathematical Formulas	68

Molar mass, and critical-point properties for a selection of gases

Substance	Formula	Molar mass (kg/kmol)	Critical-point properties	
			Temperature (K)	Pressure (bar)
Air		28.97	132.5	37.7
Ammonia	NH_3	17.03	405.5	112.8
Argon	Ar	39.95	151.0	48.6
n-Butane	C_4H_{10}	58.12	425.2	38.0
Carbon dioxide	CO_2	44.01	304.2	73.5
Carbon monoxide	CO	28.01	133.0	35.0
Chlorine	Cl_2	70.91	417.0	77.1
Ethane	C_2H_6	30.07	305.5	44.8
Ethyl alcohol	C_2H_5OH	46.07	516.0	63.8
Helium	He	4.00	5.3	2.3
Hydrogen	H_2	2.02	33.3	13.0
Methane	CH_4	16.04	191.1	46.4
Neon	Ne	20.18	44.5	27.3
Nitrogen	N_2	28.01	126.2	33.9
Nitrous oxide	N_2O	44.01	309.7	72.7
Oxygen	O_2	32.00	154.8	50.8
Propane	C_3H_8	44.10	370.0	42.6
R134a	CF_3CH_2F	102.03	374.2	40.6
R22	$CHClF_2$	86.47	369.3	49.9
Water	H_2O	18.01	647.3	221.2

Source: Cengel, Boles, 2006, **Thermodynamics**, McGraw-Hill, ISBN 0-07-288495-9

Approximative molar mass for a selection of atoms

Name	Formula	Molar mass (kg/kmol)
Hydrogen	H	1
Carbon	C	12
Nitrogen	N	14
Oxygen	O	16
Fluorine	Fl	19
Phosphorus	P	31
Sulphur	S	32
Chlorine	Cl	35.5
Bromine	Br	79.9

Thermodynamic Properties of Saturated Water H₂O

State: ' - Liquid " - Vapor

Temperature	Pressure	Specific volume		Enthalpy		Entropy	
t	p	v'	v''	h'	h''	s'	s''
°C	bar	m³/kg		kJ/kg		kJ/(kg·K)	
0.01**	0.0061124	0.0010002	206.16	0	2501.6	0	9.1575
10	0.012270	0.0010003	106.43	41.994	2519.9	0.15099	8.9020
20	0.023366	0.0010017	57.838	83.862	2538.2	0.29630	8.6684
30	0.042415	0.0010043	32.929	125.66	2556.4	0.43651	8.4546
40	0.073750	0.0010078	19.546	167.45	2574.4	0.57212	8.2583
50	0.123353	0.0010121	12.046	209.26	2592.2	0.70351	8.0776
60	0.199202	0.0010171	7.6785	251.09	2609.7	0.83099	7.9108
70	0.311620	0.0010228	5.0463	292.97	2626.9	0.95482	7.7565
80	0.473598	0.0010292	3.4091	334.92	2643.8	1.0753	7.6132
90	0.701088	0.0010361	2.3613	376.94	2660.1	1.1925	7.4799
100	1.0133	0.0010437	1.6730	419.06	2676.0	1.3069	7.3554
110	1.4327	0.0010519	1.2099	461.32	2691.3	1.4185	7.2388
120	1.9854	0.0010606	0.89152	503.72	2706.0	1.5276	7.1293
130	2.7013	0.0010700	0.66814	546.31	2719.9	1.6344	7.0261
140	3.6138	0.0010801	0.50849	589.10	2733.1	1.7390	6.9284
150	4.7600	0.0010908	0.39245	632.15	2745.4	1.8416	6.8358
160	6.1806	0.0011022	0.30676	675.47	2756.7	1.9425	6.7475
170	7.9202	0.0011145	0.24255	719.12	2767.1	2.0416	6.6630
180	10.027	0.0011275	0.19380	763.12	2776.3	2.1393	6.5819
190	12.551	0.0011415	0.15632	807.52	2784.3	2.2356	6.5036
200	15.549	0.0011565	0.12716	852.37	2790.9	2.3307	6.4278
210	19.077	0.0011726	0.10424	897.73	2796.2	2.4247	6.3539
220	23.198	0.0011900	0.086038	943.67	2799.9	2.5178	6.2817
230	27.976	0.0012087	0.071450	990.27	2802.0	2.6102	6.2107
240	33.478	0.0012291	0.059654	1037.6	2802.2	2.7020	6.1406
250	39.776	0.0012513	0.050037	1085.8	2800.4	2.7935	6.0708
260	46.943	0.0012756	0.042134	1134.9	2796.4	2.8848	6.0010
270	55.058	0.0013025	0.035588	1185.2	2789.9	2.9763	5.9304
280	64.202	0.0013324	0.030126	1236.8	2780.4	3.0683	5.8586
290	74.461	0.0013659	0.025535	1290.0	2767.6	3.1611	5.7848
300	85.927	0.0014041	0.021649	1345.1	2751.0	3.2552	5.7081
310	98.700	0.0014480	0.018334	1402.4	2730.0	3.3512	5.6278
320	112.89	0.0014995	0.015480	1462.6	2703.7	3.4500	5.5423
330	128.63	0.0015615	0.012989	1526.5	2670.2	3.5528	5.4490
340	146.05	0.0016387	0.010780	1595.5	2626.2	3.6616	5.3427
350	165.35	0.0017411	0.0087991	1671.9	2567.7	3.7800	5.2177
352	169.45	0.0017661	0.0084205	1689.3	2553.5	3.8071	5.1893
354	173.64	0.0017937	0.0080453	1707.5	2538.4	3.8349	5.1596
356	177.92	0.0018241	0.0076741	1725.9	2522.1	3.8629	5.1283
358	182.29	0.0018580	0.0073061	1744.7	2504.6	3.8915	5.0953
360	186.75	0.0018959	0.0069398	1764.2	2485.4	3.9210	5.0600
362	191.31	0.0019388	0.0065727	1784.6	2464.4	3.9518	5.0220
364	195.96	0.0019882	0.0062010	1806.4	2440.9	3.9846	4.9804
366	200.72	0.0020464	0.0058186	1830.2	2414.1	4.0205	4.9339
368	205.57	0.0021181	0.0054157	1857.3	2382.4	4.0613	4.8801
370	210.54	0.0022136	0.0049727	1890.2	2342.8	4.1108	4.8144
372	215.62	0.0023636	0.0044389	1935.6	2287.0	4.1794	4.7240
374	220.81	0.0028427	0.0034659	2046.7	2156.2	4.3493	4.5185
374.15*	221.20	0.0031700		2107.4		4.4429	

** - Triple point
* - Critical point

Based on the Fundamental Equation of State developed by E. Schmidt
Properties of Water and Steam in SI-Units, Second Edition
Springer-Verlag Berlin Heidelberg and R. Oldenbourg, München 1979
(In accordance with the International Skeleton Tables ratified by the Sixth International Conference on the
Properties of Steam, New York, October 1963)

Thermodynamic Properties of Saturated Nitrogen N_2

State:　　' - Liquid　　" - Vapor

Temperature	Pressure	Specific volume		Enthalpy		Entropy	
T	p	v'	v''	h'	h''	s'	s''
K	bar	m^3/kg		kJ/kg		kJ/(kg·K)	
63.148**	0.12520	0.001150	1.481	-460.3	-244.9	-4.412	-1.001
64	0.14600	0.001155	1.286	-458.6	-244.1	-4.385	-1.033
66	0.20630	0.001166	0.9354	-454.5	-242.3	-4.323	-1.106
68	0.28500	0.001178	0.6949	-450.5	-240.4	-4.263	-1.174
70	0.38570	0.001191	0.5262	-446.5	-238.7	-4.204	-1.236
72	0.51250	0.001203	0.4053	-442.4	-237.0	-4.148	-1.294
74	0.66960	0.001217	0.3169	-438.4	-235.3	-4.092	-1.348
76	0.86170	0.001230	0.2513	-434.3	-233.7	-4.038	-1.399
78	1.0935	0.001244	0.2017	-430.2	-232.2	-3.986	-1.447
80	1.3699	0.001259	0.1637	-426.1	-230.8	-3.934	-1.492
82	1.6961	0.001274	0.1343	-421.9	-229.4	-3.883	-1.535
84	2.0773	0.001290	0.1112	-417.7	-228.1	-3.833	-1.576
86	2.5191	0.001307	0.09284	-413.5	-226.9	-3.784	-1.614
88	3.0270	0.001325	0.07810	-409.3	-225.8	-3.736	-1.651
90	3.6066	0.001343	0.06615	-405.0	-224.8	-3.689	-1.687
92	4.2639	0.001362	0.05637	-400.6	-224.0	-3.642	-1.722
94	5.0047	0.001383	0.04830	-396.2	-223.2	-3.596	-1.755
96	5.8349	0.001404	0.04159	-391.7	-222.6	-3.550	-1.788
98	6.7607	0.001427	0.03596	-387.2	-222.2	-3.504	-1.820
100	7.7881	0.001452	0.03122	-382.6	-221.9	-3.459	-1.852
102	8.9234	0.001478	0.02718	-377.9	-221.8	-3.414	-1.884
104	10.173	0.001507	0.02373	-373.1	-221.9	-3.370	-1.916
106	11.543	0.001538	0.02077	-368.1	-222.2	-3.325	-1.948
108	13.040	0.001572	0.01819	-363.1	-222.8	-3.279	-1.981
110	14.672	0.001610	0.01595	-357.8	-223.7	-3.234	-2.014
112	16.445	0.001653	0.01398	-352.4	-224.9	-3.187	-2.049
114	18.368	0.001701	0.01224	-346.7	-226.5	-3.140	-2.085
116	20.448	0.001759	0.01069	-340.6	-228.6	-3.090	-2.124
118	22.697	0.001828	0.009286	-334.1	-231.4	-3.038	-2.167
120	25.125	0.001915	0.007998	-327.0	-235.1	-2.982	-2.216
122	27.747	0.002032	0.006783	-318.9	-240.1	-2.919	-2.274
124	30.582	0.002209	0.005563	-309.1	-247.8	-2.845	-2.350
126	33.665	0.002683	0.003893	-291.7	-266.7	-2.711	-2.513
126.193*	33.978	0.003202		-279.4		-2.615	

** - Triple point
* - Critical point

Based on the Fundamental Equation of State developed by:
R.T. Jacobsen, R.B. Stewart, and M. Jahangiri
Thermodynamic Properties of Nitrogen from the Freezing Line to 2000 K at Pressures to 1000 MPa
J. Phys, Chem, Ref. Data, Vol. 15, No. 2, 1986.

Thermodynamic Properties of Saturated Carbon Dioxide CO$_2$

State:	' - Solid	" - Vapor					
Temperature	Pressure	Specific volume		Enthalpy		Entropy	
t	p	v'	v"	h'	h"	s'	s"
°C	bar	m³/kg		kJ/kg		kJ/(kg·K)	
-100	0.139	0.000627	2.336	-173.1	412.1	-0.676	2.704
-95	0.231	0.000629	1.442	-167.5	415.3	-0.644	2.628
-90	0.372	0.000632	0.9201	-161.8	418.4	-0.612	2.557
-85	0.584	0.000635	0.5981	-156.0	421.3	-0.581	2.488
-80	0.896	0.000639	0.3979	-150.0	424.0	-0.550	2.423
-75	1.35	0.000643	0.2695	-143.6	426.3	-0.517	2.360
-70	1.99	0.000647	0.1854	-136.7	428.3	-0.483	2.299
-65	2.88	0.000652	0.1293	-128.7	429.7	-0.444	2.239
-60	4.10	0.000657	0.09115	-119.4	430.5	-0.400	2.180
-56.55**	5.18	0.000661	0.07222	-113.1	430.7	-0.371	2.140

** - Triple point

Based on data from Plank, R., Kuprianoff, J. Z. ges. Kälteind., Vol. 36,
s. 41 and Handbuch der Kältechnik, Bd 4 (1956), by R. Plank

State:	' - Liquid	" - Vapor					
Temperature	Pressure	Specific volume		Enthalpy		Entropy	
t	p	v'	v"	h'	h"	s'	s"
°C	bar	m³/kg		kJ/kg		kJ/(kg·K)	
-56.55**	5.181	0.000849	0.07264	80.05	430.4	0.5214	2.139
-56	5.306	0.000850	0.07101	81.13	430.6	0.5263	2.136
-54	5.780	0.000855	0.06543	85.06	431.3	0.5441	2.124
-52	6.286	0.000861	0.06037	89.00	432.0	0.5618	2.113
-50	6.823	0.000866	0.05579	92.94	432.7	0.5794	2.102
-48	7.395	0.000872	0.05162	96.91	433.3	0.5968	2.091
-46	8.002	0.000878	0.04782	100.9	433.9	0.6142	2.080
-44	8.645	0.000883	0.04435	104.9	434.4	0.6314	2.069
-42	9.325	0.000889	0.04118	108.9	434.9	0.6486	2.059
-40	10.045	0.000896	0.03828	112.9	435.3	0.6656	2.049
-38	10.805	0.000902	0.03562	116.9	435.7	0.6826	2.038
-36	11.607	0.000909	0.03318	121.0	436.1	0.6995	2.028
-34	12.453	0.000915	0.03093	125.1	436.4	0.7163	2.018
-32	13.342	0.000922	0.02886	129.2	436.6	0.7331	2.008
-30	14.278	0.000930	0.02695	133.3	436.8	0.7498	1.998
-28	15.261	0.000937	0.02519	137.5	437.0	0.7665	1.988
-26	16.293	0.000945	0.02356	141.7	437.0	0.7831	1.978
-24	17.375	0.000953	0.02205	145.9	437.1	0.7997	1.968
-22	18.509	0.000961	0.02065	150.2	437.0	0.8163	1.958
-20	19.696	0.000969	0.01934	154.4	436.9	0.8328	1.949
-18	20.938	0.000978	0.01813	158.8	436.7	0.8494	1.939
-16	22.236	0.000987	0.01700	163.1	436.4	0.8659	1.929
-14	23.593	0.000997	0.01595	167.6	436.1	0.8825	1.919
-12	25.009	0.001007	0.01497	172.0	435.7	0.8991	1.909
-10	26.486	0.001017	0.01405	176.5	435.1	0.9157	1.899
-8	28.026	0.001028	0.01319	181.1	434.5	0.9324	1.888
-6	29.631	0.001040	0.01238	185.7	433.8	0.9491	1.878
-4	31.302	0.001052	0.01162	190.4	432.9	0.9660	1.867
-2	33.041	0.001065	0.01091	195.2	432.0	0.9829	1.856
0	34.850	0.001078	0.01024	200.0	430.9	1.000	1.845
2	36.732	0.001093	0.009609	204.9	429.6	1.017	1.834
4	38.688	0.001108	0.009011	210.0	428.2	1.035	1.822
6	40.720	0.001124	0.008445	215.1	426.7	1.052	1.810
8	42.831	0.001142	0.007909	220.3	424.9	1.070	1.798
10	45.022	0.001161	0.007398	225.7	422.9	1.088	1.785
12	47.297	0.001182	0.006912	231.3	420.6	1.107	1.771
14	49.659	0.001205	0.006447	237.0	418.0	1.126	1.757
16	52.109	0.001231	0.006000	243.0	415.1	1.146	1.741
18	54.653	0.001260	0.005568	249.2	411.8	1.166	1.724
20	57.292	0.001293	0.005149	255.8	407.9	1.188	1.706
22	60.032	0.001331	0.004738	262.8	403.3	1.210	1.686
24	62.879	0.001378	0.004328	270.4	397.7	1.235	1.663
26	65.837	0.001436	0.003912	278.9	390.8	1.262	1.636
28	68.918	0.001517	0.003467	288.8	381.5	1.293	1.601
30	72.136	0.001660	0.002919	302.8	366.0	1.338	1.546
30,98*	73.773	0.002174		334.1		1.440	

** - Triple point
* - Critical point

Based on the Fundamental Equation of State developed by:
R. Span and W. Wagner, A New Equation of State for Carbon Dioxide Covering the Fluid Region form
the Triple-Point Temperature to 1100 K at Pressures up to 800 MPa,
J. Phys. Chem, Ref. Data, Vol. 25, No. 6, 1996.

Thermodynamic Properties of Saturated Ammonia, NH$_3$

State: ' - Liquid, " - Vapor

Temperature	Pressure	Specific volume		Enthalpy		Entropy	
t	p	v'	v''	h'	h''	s'	s''
°C	bar	m³/kg		kJ/kg		kJ/(kg·K)	
-50	0.4082	0.001424	2.631	-24.7276	1391.3	0.09450	6.440
-45	0.5447	0.001436	2.008	-2.8474	1399.6	0.1914	6.339
-40	0.7166	0.001449	1.553	19.1701	1407.7	0.2867	6.242
-35	0.9307	0.001462	1.215	41.32	1415.5	0.3806	6.151
-30	1.194	0.001475	0.9619	63.60	1423.0	0.4730	6.064
-25	1.515	0.001489	0.7717	86.01	1430.7	0.5641	5.983
-20	1.901	0.001503	0.6237	108.6	1437.7	0.6538	5.904
-15	2.362	0.001518	0.5086	131.2	1444.4	0.7421	5.829
-10	2.908	0.001534	0.4182	154.0	1450.7	0.8293	5.757
-5	3.549	0.001550	0.3465	176.9	1456.7	0.9152	5.687
0	4.296	0.001566	0.2892	200.0	1462.2	1.000	5.621
5	5.160	0.001583	0.2429	223.2	1467.4	1.084	5.557
10	6.153	0.001601	0.2053	246.6	1472.1	1.166	5.494
15	7.288	0.001619	0.1745	270.1	1476.4	1.248	5.434
20	8.578	0.001639	0.1491	293.8	1480.1	1.329	5.376
25	10.03	0.001659	0.1279	317.7	1483.2	1.409	5.318
30	11.67	0.001680	0.1104	341.8	1486.2	1.488	5.263
35	13.51	0.001702	0.09563	366.1	1488.3	1.567	5.209
40	15.55	0.001726	0.08311	390.6	1489.9	1.645	5.155
45	17.82	0.001750	0.07247	415.5	1490.9	1.722	5.102
50	20.33	0.001777	0.06337	440.6	1491.1	1.799	5.050

Data from EES V8.429-3D (2009).

Thermodynamic Properties of Saturated R22

State: ' - Liquid, " - Vapor

Temperature	Pressure	Specific volume		Enthalpy		Entropy	
t	p	v'	v''	h'	h''	s'	s''
°C	bar	m³/kg		kJ/kg		kJ/(kg·K)	
-50	0.6452	0.0006967	0.3236	143.6	383.2	0.7734	1.847
-45	0.8291	0.0007037	0.2562	149.0	385.6	0.7976	1.834
-40	1.052	0.0007110	0.2051	154.5	387.9	0.8213	1.822
-35	1.320	0.0007185	0.1659	160.1	390.2	0.8447	1.811
-30	1.639	0.0007263	0.1354	165.6	392.5	0.8678	1.801
-25	2.015	0.0007343	0.1116	171.2	394.7	0.8905	1.791
-20	2.454	0.0007427	0.09263	176.9	396.9	0.9129	1.782
-15	2.963	0.0007515	0.07747	182.6	399.0	0.9350	1.774
-10	3.549	0.0007607	0.06523	188.3	401.1	0.9569	1.765
-5	4.219	0.0007702	0.05526	194.1	403.1	0.9786	1.758
0	4.981	0.0007803	0.04708	200.0	405.0	1.000	1.750
5	5.843	0.0007908	0.04032	205.9	406.8	1.021	1.743
10	6.812	0.0008020	0.03468	211.9	408.5	1.042	1.737
15	7.896	0.0008137	0.02996	218.0	410.1	1.063	1.730
20	9.103	0.0008262	0.02598	224.2	411.7	1.084	1.724
25	10.44	0.0008395	0.02260	230.4	413.0	1.105	1.717
30	11.92	0.0008538	0.01971	236.8	414.3	1.126	1.711
35	13.55	0.0008691	0.01724	243.2	415.4	1.146	1.705
40	15.34	0.0008857	0.01510	249.8	416.3	1.167	1.699
45	17.30	0.0009038	0.01324	256.5	417.0	1.188	1.692
50	19.43	0.0009236	0.01163	263.4	417.5	1.209	1.685

Data from EES V8.429-3D (2009).

Thermal properties of water

Critical point : $p_k = 221.2$ bar, $T_k = 647.3$ K, $v_k = 0.003170$ m³/kg

Solid water (ice)

Temperature	Density	Specific heat	Thermal conductivity
t	ρ	c_p	λ
°C	kg/m³	kJ/(kg·K)	W/(m·K)
-80	920	2.114	2.23
-40	920	1.805	2.44
-20	920	1.574	2.66
0	920	1.352	3.17

Liquid water

Temperature	Density	Specific heat	Thermal conductivity	Kinematic viscosity	Dynamic viscosity	Prandtl Number	Rayleigh Function
t	ρ	c_p	λ	v	μ	Pr	C_{Ra}
°C	kg/m³	kJ/(kg·K)	W/(m·K)	m²/s	Pa·s	-	1/(m³·K)
0	1000	4.205	0.564	1.79E-06	1.79E-03	13.4	-2.10E+08
7	1000	4.297	0.582	1.44E-06	1.44E-03	10.4	2.20E+08
27	997	4.177	0.608	8.60E-07	8.57E-04	5.88	3.66E+09
47	989	4.176	0.637	5.85E-07	5.79E-04	3.79	1.25E+10
67	980	4.187	0.659	4.32E-07	4.23E-04	2.69	2.98E+10
87	967	4.204	0.674	3.31E-07	3.20E-04	2.00	6.25E+10
100	958	4.220	0.681	2.94E-07	2.82E-04	1.75	8.50E+10
127	937	4.241	0.686	2.34E-07	2.19E-04	1.35	1.61E+11
177	890	4.419	0.673	1.72E-07	1.53E-04	1.01	4.02E+11
227	832	4.647	0.635	1.42E-07	1.18E-04	0.86	7.71E+11
277	756	5.272	0.571	1.26E-07	9.50E-05	0.88	1.44E+12
327	650	6.691	0.481	1.17E-07	7.60E-05	1.05	2.95E+12

Water vapor at atmospheric pressure (1.013 bar)

Temperature	Density	Specific heat	Thermal conductivity	Kinematic viscosity	Dynamic viscosity	Prandtl Number	Rayleigh Function
t	ρ	c_p	λ	v	μ	Pr	C_{Ra}
°C	kg/m³	kJ/(kg·K)	W/(m·K)	m²/s	Pa·s	-	1/(m³·K)
127	0.555	2.000	0.0264	2.38E-05	1.32E-05	1.00	4.33E+07
177	0.491	1.968	0.0307	3.10E-05	1.52E-05	0.98	2.27E+07
227	0.441	1.977	0.0657	3.92E-05	1.73E-05	0.96	1.28E+07
277	0.401	1.994	0.0411	4.81E-05	1.93E-05	0.94	7.70E+06
327	0.367	2.022	0.0464	5.80E-05	2.13E-05	0.93	4.80E+06
427	0.314	2.083	0.0572	8.09E-05	2.54E-05	0.93	2.14E+06
527	0.275	2.148	0.0686	1.07E-04	2.95E-05	0.92	1.06E+06
627	0.244	2.217	0.0780	1.38E-04	3.36E-05	0.95	5.80E+05
727	0.220	2.288	0.0870	1.71E-04	3.76E-05	0.99	3.30E+05

Source: Granryd, E., 2003, **Heat Transfer – Collection of formulas**, KTH Energy Technology.

Thermal properties of air

Composition by volume : 78.03% N_2, 20.99% O_2, 0.933% Ar, 0.030% CO_2, 0.01% H_2

Critical point : $p_k = 37.7$ bar, $T_k = 132.5$ K, $v_k = 0.003048$ m³/kg

Air at atmospheric pressure (1.013 bar)

Temperature	Density	Specific heat	Thermal conductivity	Kinematic viscosity	Dynamic viscosity	Prandtl Number	Rayleigh Function
t	ρ	c_p	λ	v	μ	Pr	C_{Ra}
°C	kg/m³	kJ/(kg·K)	W/(m·K)	m²/s	Pa·s	-	1/(m³·K)
-150	2.897	1.016	0.0113	2.93E-06	8.50E-06	7.64E-01	7.072E+09
-100	2.046	1.008	0.0158	5.72E-06	1.17E-05	7.46E-01	1.293E+09
-50	1.584	1.006	0.0200	9.22E-06	1.46E-05	7.34E-01	3.800E+08
0	1.293	1.006	0.0241	1.33E-05	1.72E-05	7.18E-01	1.457E+08
20	1.205	1.006	0.0257	1.50E-05	1.81E-05	7.09E-01	1.051E+08
40	1.127	1.007	0.0273	1.69E-05	1.91E-05	7.05E-01	7.684E+07
60	1.060	1.008	0.0288	1.89E-05	2.00E-05	7.00E-01	5.790E+07
80	1.000	1.010	0.0303	2.09E-05	2.09E-05	6.97E-01	4.430E+07
100	0.946	1.011	0.0317	2.29E-05	2.17E-05	6.92E-01	3.458E+07
150	0.834	1.017	0.0353	2.85E-05	2.38E-05	6.86E-01	1.952E+07
200	0.746	1.025	0.0387	3.45E-05	2.57E-05	6.81E-01	1.189E+07
300	0.616	1.045	0.0450	4.76E-05	2.93E-05	6.80E-01	5.148E+06
400	0.524	1.069	0.0509	6.20E-05	3.25E-05	6.83E-01	2.586E+06
500	0.456	1.093	0.0563	7.79E-05	3.55E-05	6.89E-01	1.443E+06
750	0.345	1.144		1.22E-04	4.21E-05		
1000	0.277	1.193		1.73E-04	4.79E-05		
1500	0.199	1.282		2.91E-04	5.79E-05		
2000	0.155	1.468					
2500	0.126	2.200					

Between 0 °C and 100 °C the following equations can be used

$C_{Ra} = Gr \cdot Pr/(\Delta t \cdot H^3) = -44.11477 \cdot t^3 + 16279.84 \cdot t^2 - 2289826 \cdot t + 1.455518 \cdot 10^8$ [1/(K·m³)]

$\lambda = 8.092374 \cdot 10^{-5} \cdot t + 2.409104 \cdot 10^{-2}$ [W/(m·K)]

$Pr = 1.728523 \cdot 10^{-8} \cdot t^3 - 2.386791 \cdot 10^{-7} \cdot t^2 - 3.541834 \cdot 10^{-4} \cdot t + 0.7180716$ [–]

$v = 8.678508 \cdot 10^{-8} \cdot t + 1.333458 \cdot 10^{-5}$ [m²/s]

t should be inserted in °C in the above equations!

Air at pressure p = 10.13 bar

Temperature	Density	Specific heat
t	ρ	c_p
°C	kg/m³	kJ/(kg·K)
-100	21.21	1.067
-50	16.06	1.035
0	13.00	1.023
50	10.99	1.019
100	9.45	1.020
150	8.32	1.024
250	6.72	1.039
500	4.55	1.094
1000	2.77	1.193
1500	1.99	1.280
2000	1.55	1.395
2500	1.27	1.678

Source: Granryd, E., 2003, **Heat Transfer – Collection of formulas**, KTH Energy Technology.

Emissivity (and Absorptivity) for various surfaces

	Emissivity ε (Absorptivity a) for	
	approx 30 °C	Solar radiation
Aluminium, polished surface	0.05	0.3
Aluminium, eloxated surface	0.8	0.16
Concrete, Red bricks	0.9	0.6
Asphalt	0.9	0.9
Wood	0.85	0.35
Roof Shingles	0.9	0.9
Paint, white	0.95	0.2
Paint, black	0.95	0.95
Water, Ice, Frost	0.95	0.2 – 0.7
Snow	0.97	0.28
Human skin (Caucasian)	0.97	0.62

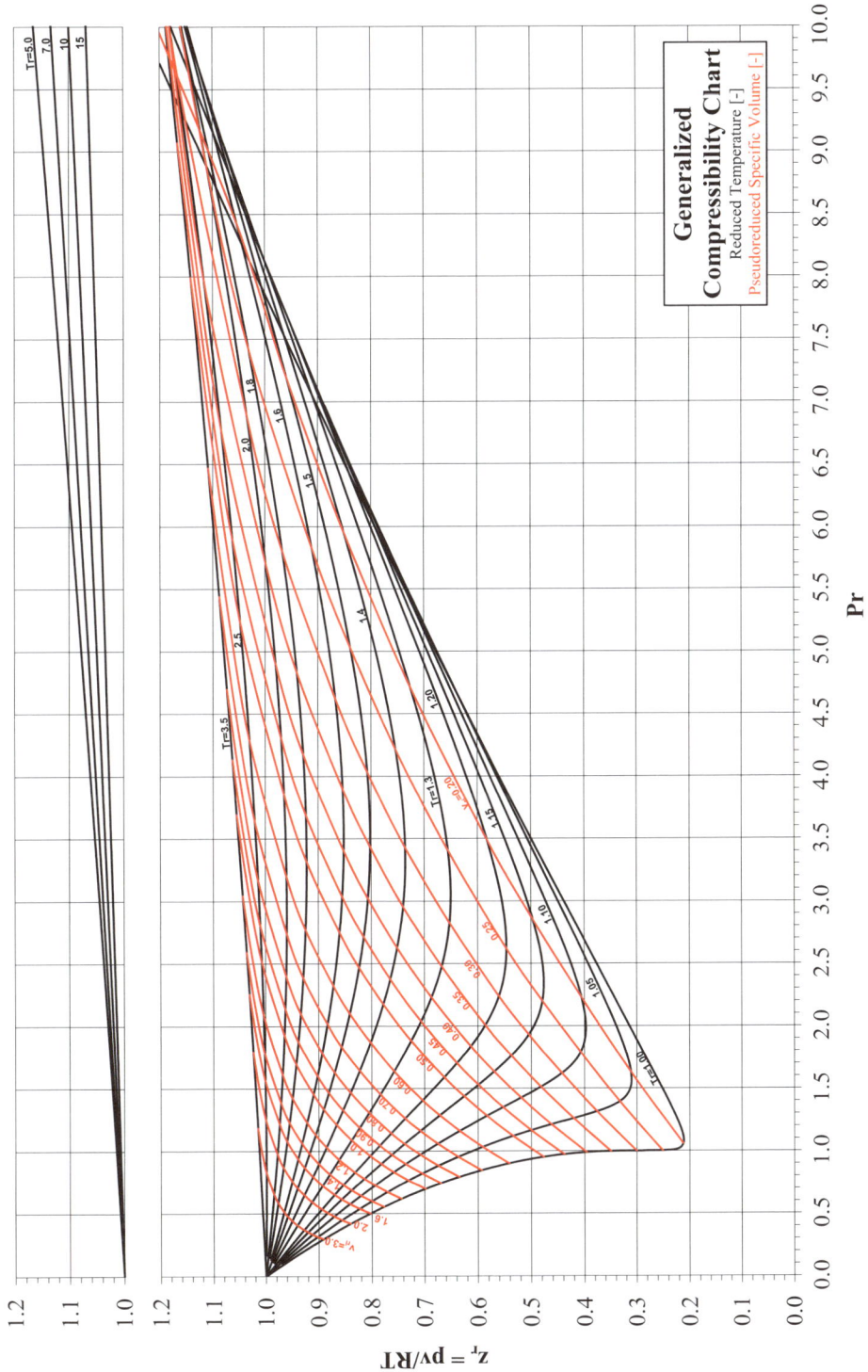

Generalized Compressibility Chart
Reduced Temperature [-]
Pseudoreduced Specific Volume [-]

$z_r = pv/RT$

P_r

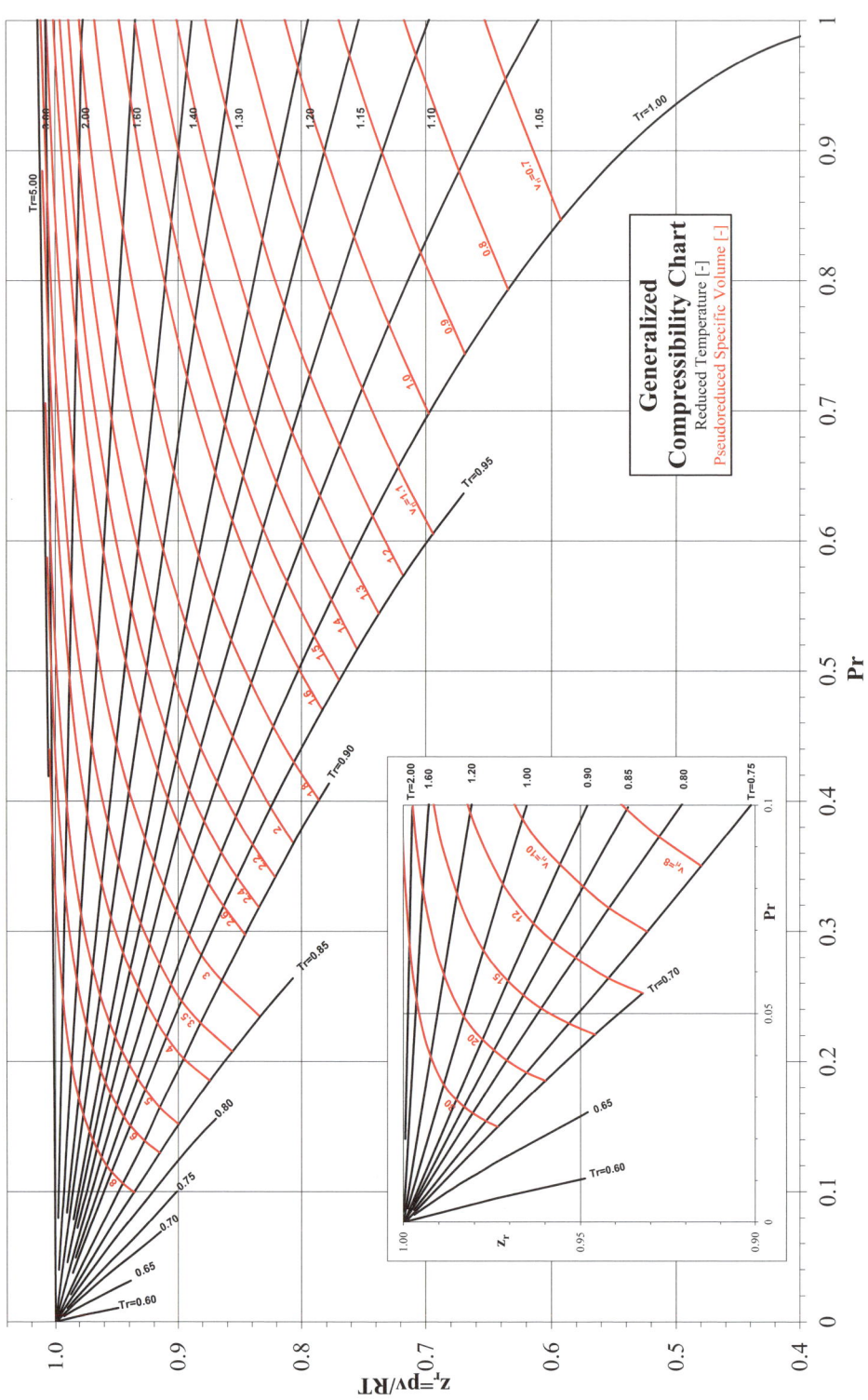

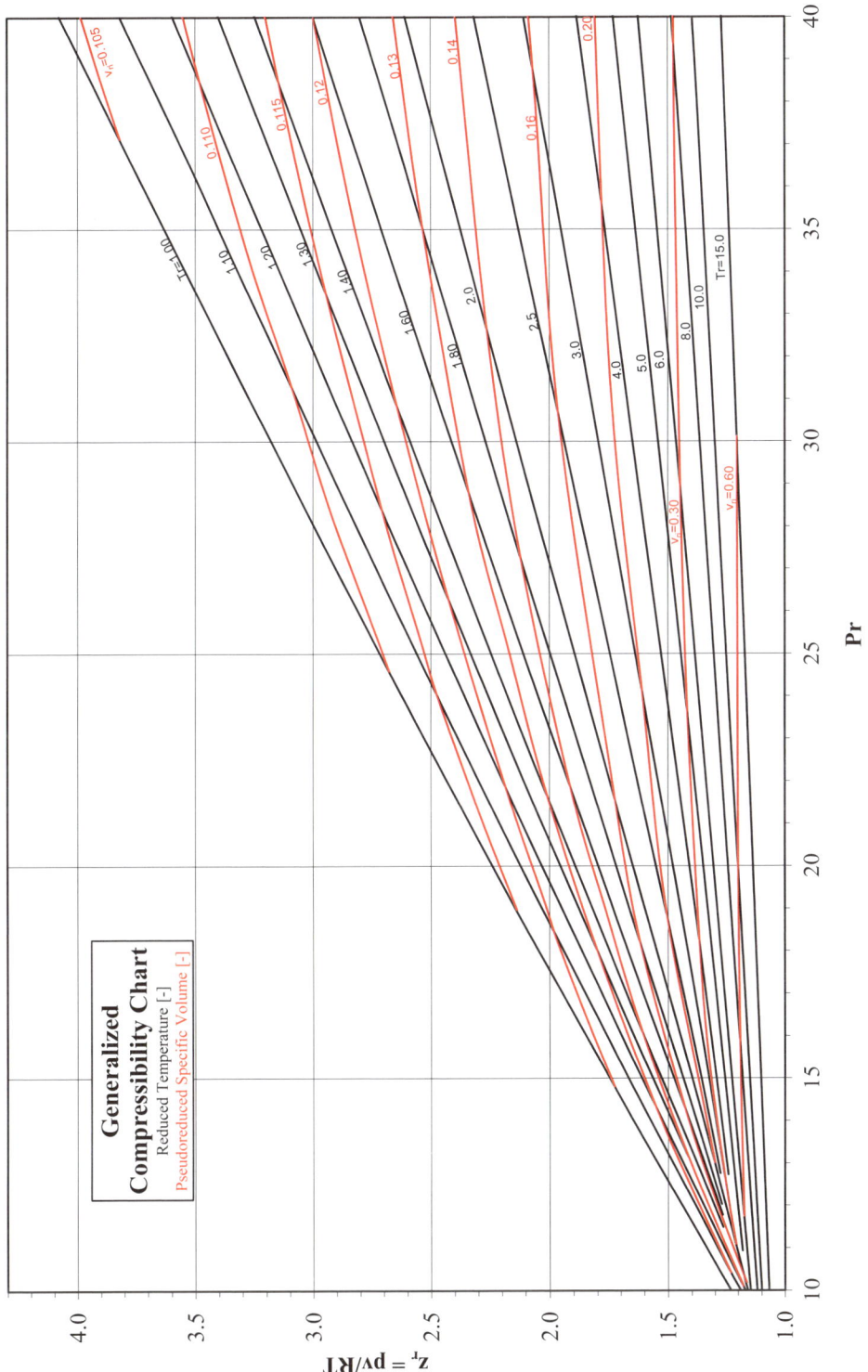

Generalized Compressibility Chart
Reduced Temperature [-]
Pseudoreduced Specific Volume [-]

$z_r = pv/RT$

Pr

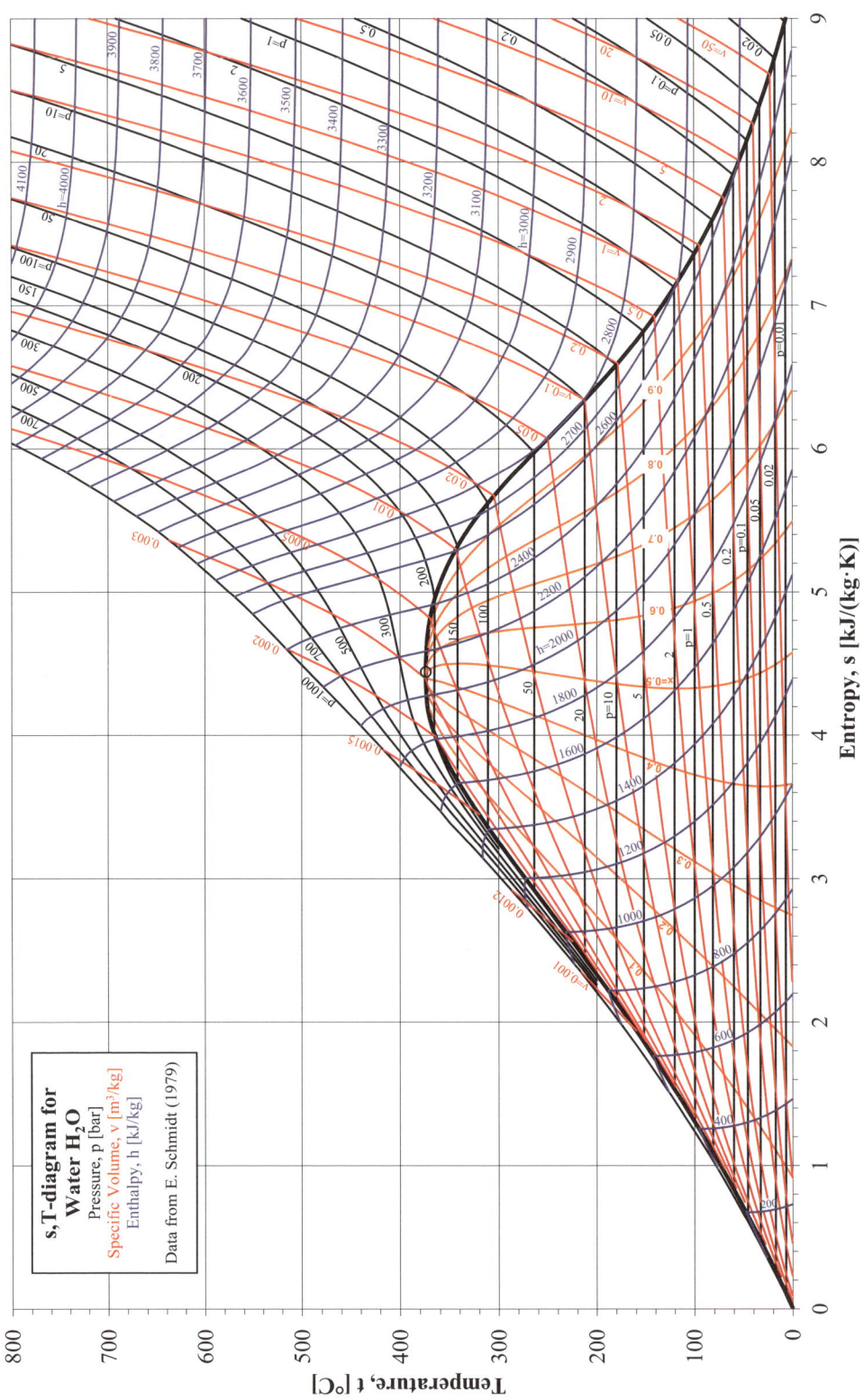

s,T-diagram for Water H$_2$O

Pressure, p [bar]
Specific Volume, v [m^3/kg]
Enthalpy, h [kJ/kg]

Data from E. Schmidt (1979)

Temperature, t [°C]

Entropy, s [kJ/(kg·K)]

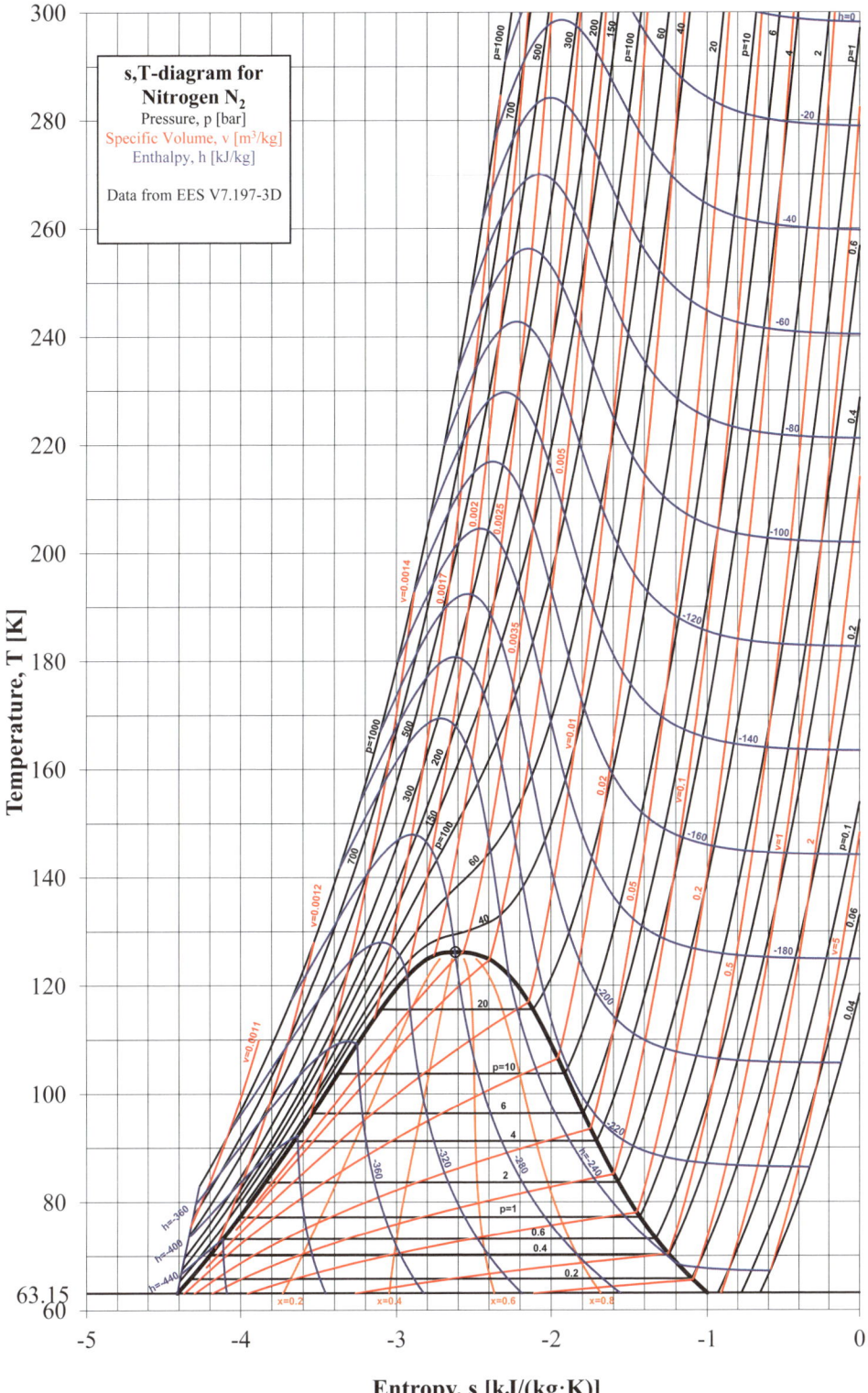

s,T-diagram for Nitrogen N₂
Pressure, p [bar]
Specific Volume, v [m³/kg]
Enthalpy, h [kJ/kg]

Data from EES V7.197-3D

Temperature, T [K]

Entropy, s [kJ/(kg·K)]

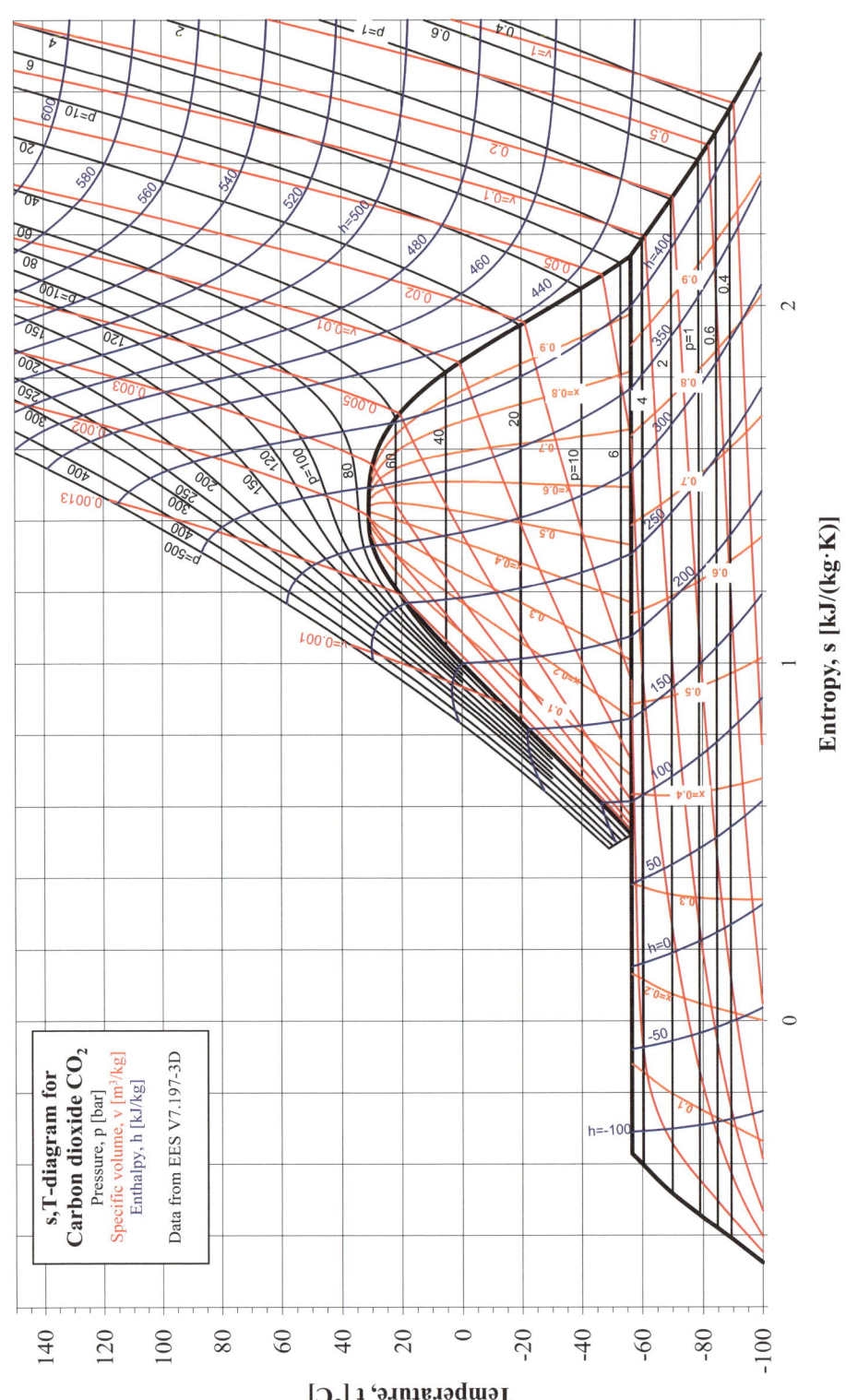

s,T-diagram for
Carbon dioxide CO₂
Pressure, p [bar]
Specific volume, v [m³/kg]
Enthalpy, h [kJ/kg]

Data from EES V7.197-3D

Temperature, t [°C]

Entropy, s [kJ/(kg·K)]

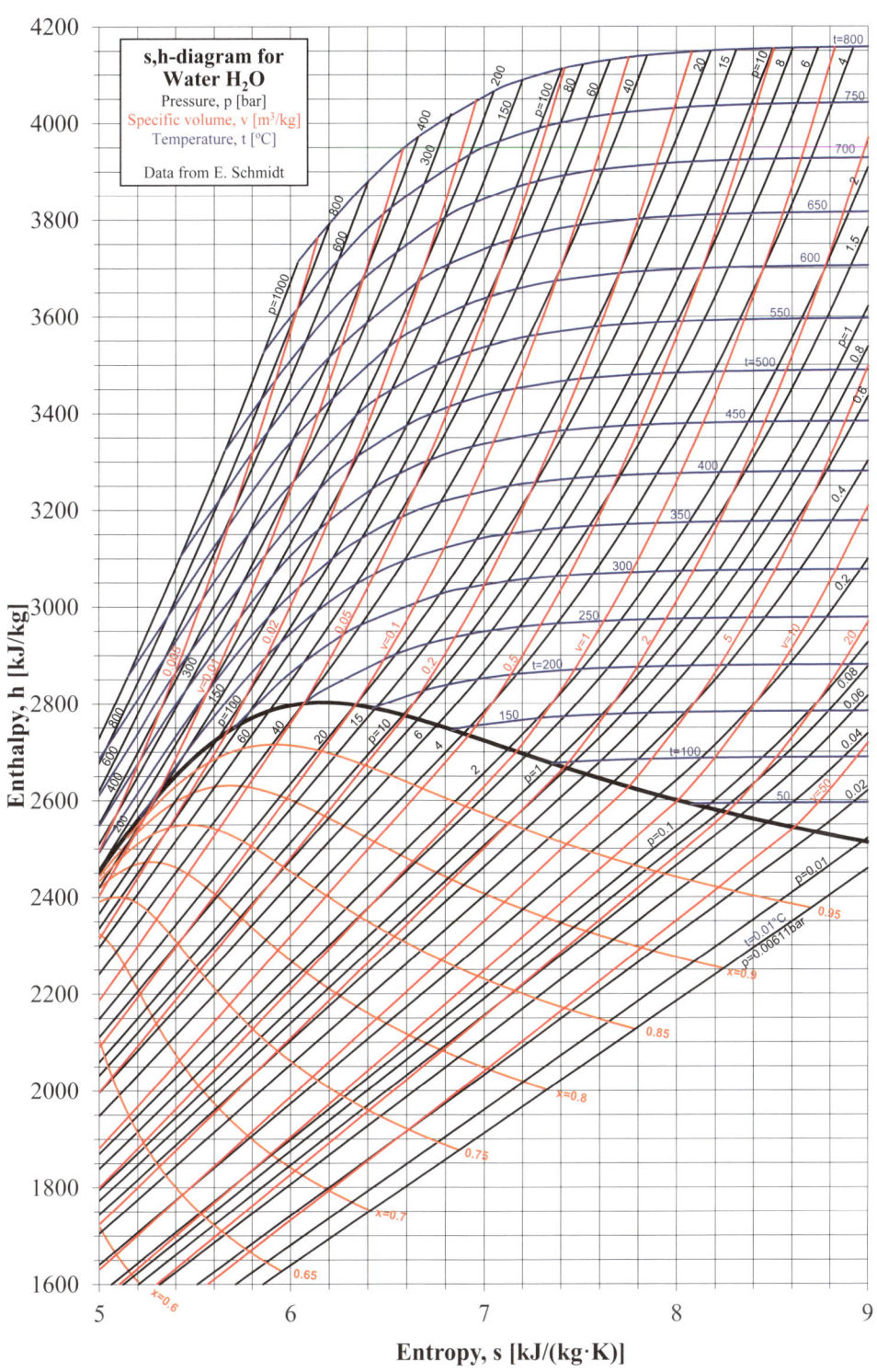

s,h-diagram for
Water H₂O

Pressure, p [bar]
Specific volume, v [m³/kg]
Temperature, t [°C]

Data from E. Schmidt

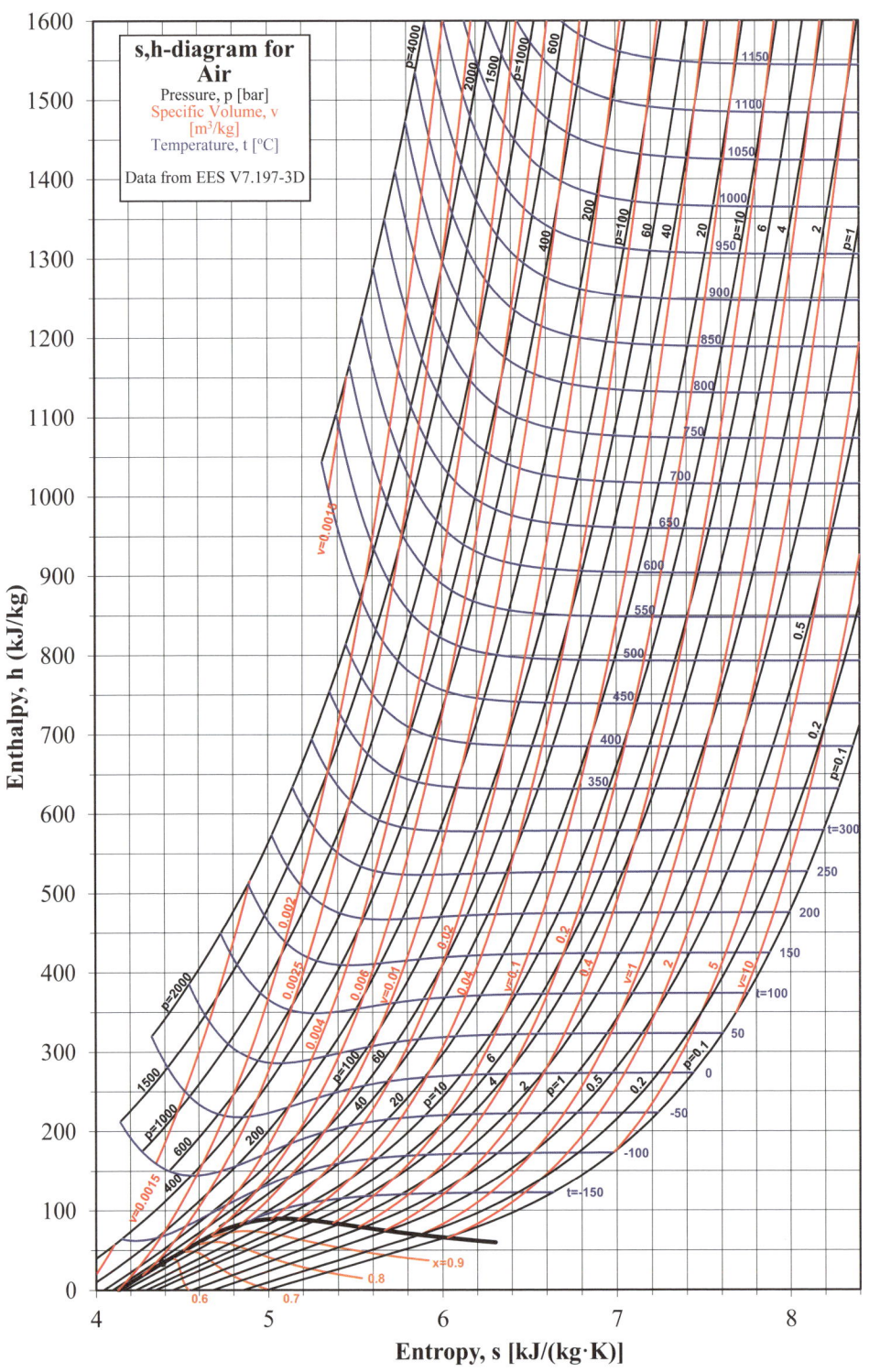

s,h-diagram for Air
Pressure, p [bar]
Specific Volume, v [m³/kg]
Temperature, t [°C]

Data from EES V7.197-3D

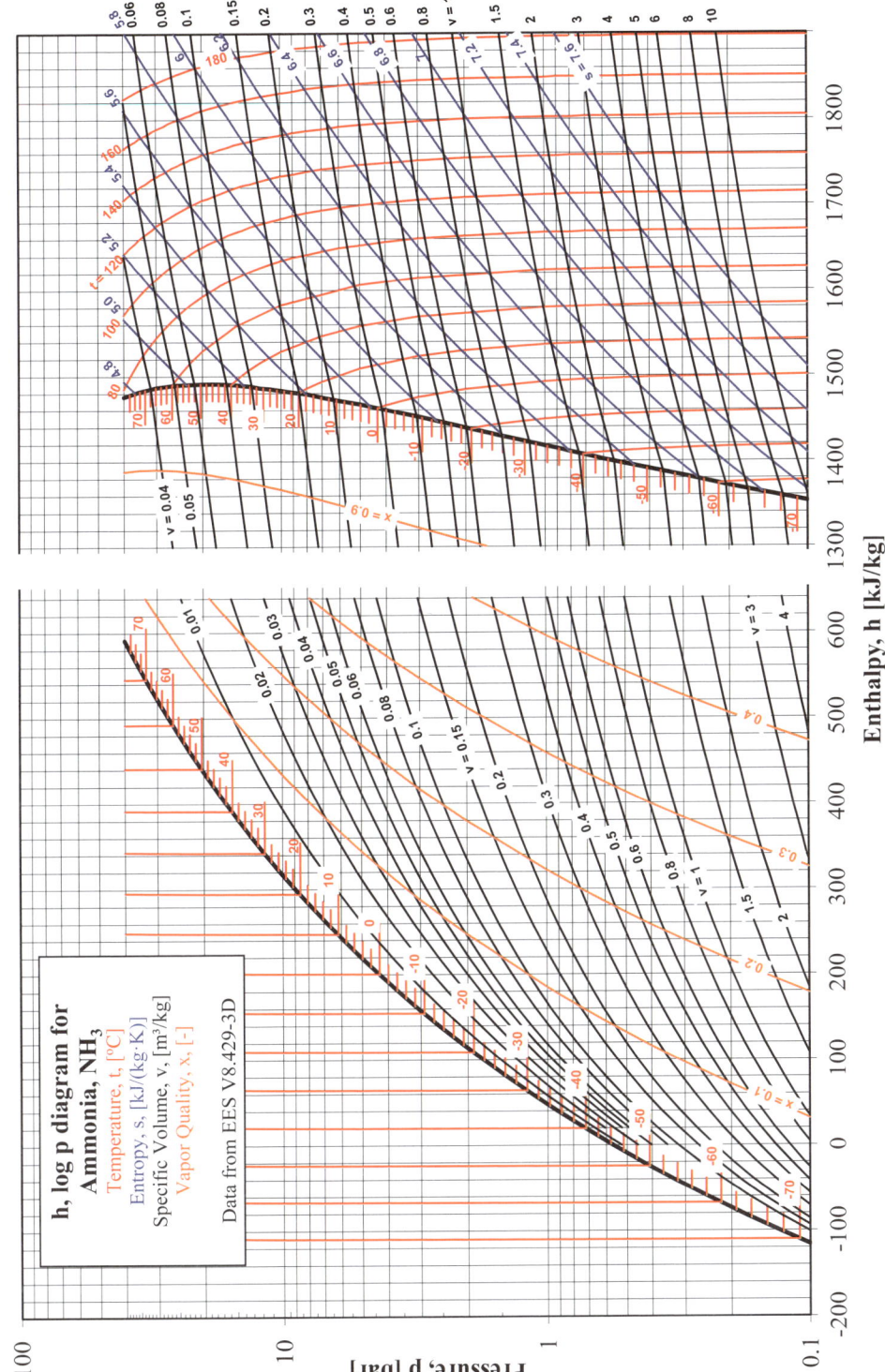

h, log p diagram for Ammonia, NH₃

Temperature, t, [°C]
Entropy, s, [kJ/(kg·K)]
Specific Volume, v, [m³/kg]
Vapor Quality, x, [-]

Data from EES V8.429-3D

Pressure, p [bar]

Enthalpy, h [kJ/kg]

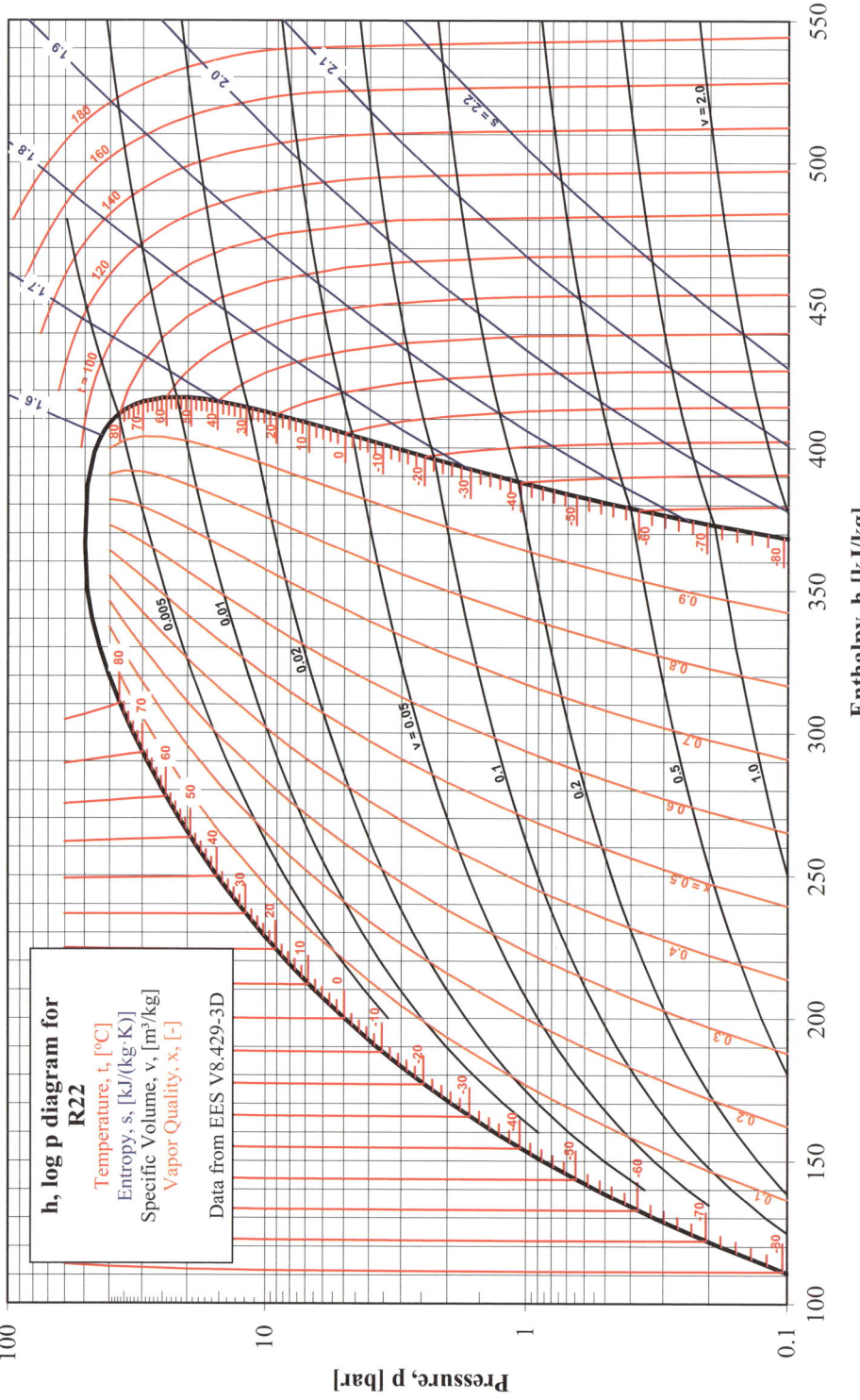

h, log p diagram for R22

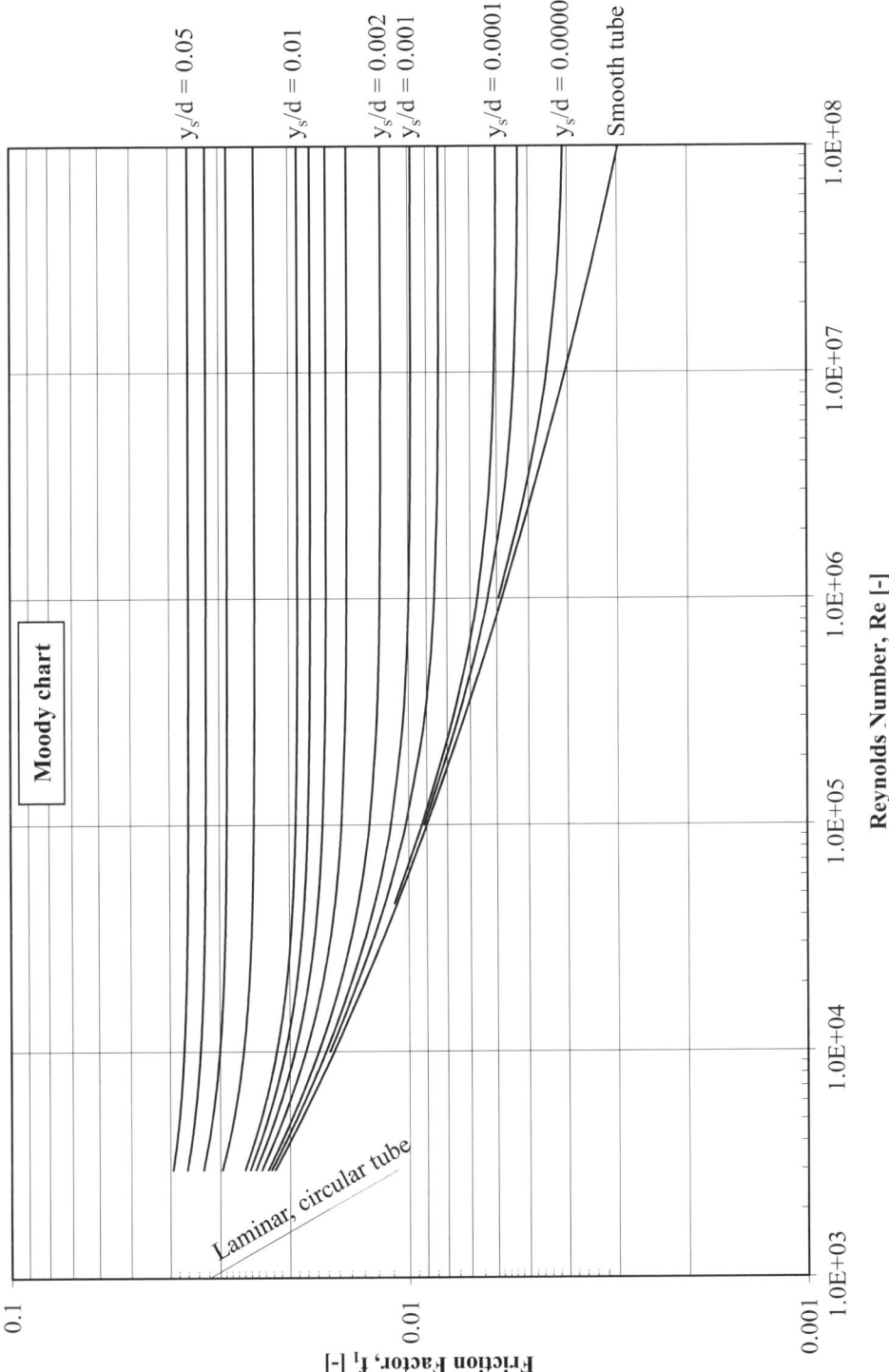

Pressure loss due to sudden area change

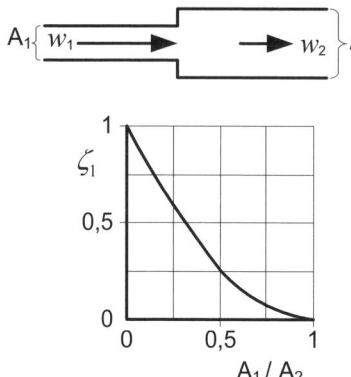

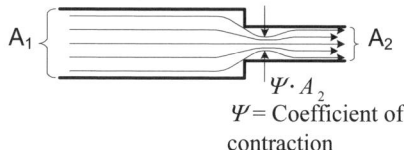

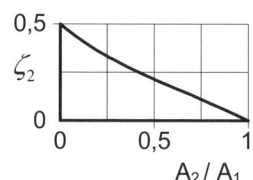

Ψ = Coefficient of contraction

Pressure loss at inlets

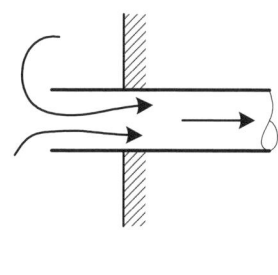

$\zeta \approx 1$

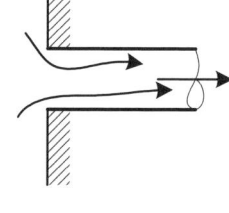

$\zeta \approx 0{,}5$

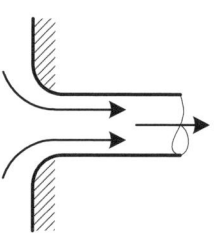

$\zeta \approx 0{,}05$

Pressure loss in valves

Disk valve	Gate valve	Check valve

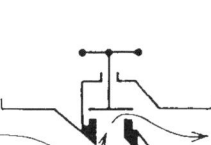

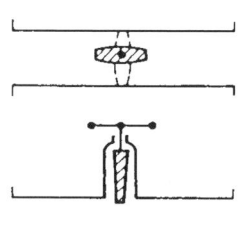

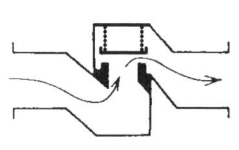

$\zeta = 2\text{–}7$

$\zeta = 0.3\text{–}1$

$\zeta = 1\text{–}7$

Pressure loss in bends, and knees

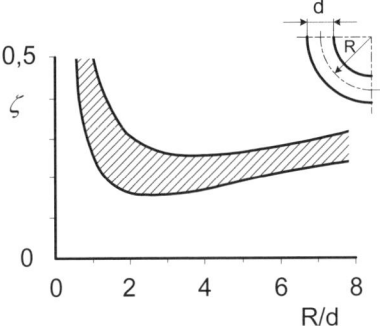

Bends

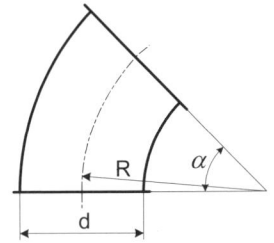

	ς		
R / d	$\alpha = 15°$	45°	90°
0.7	-	-	0.66
1	0.05	0.15	0.30
3	0.03	0.09	0.13
5	0.03	0.08	0.10
10	0.03	0.07	0.07

Knees

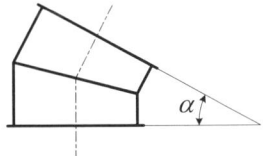

$\alpha = 20°$	60°	80°	90°	120°	140°	160°
$\varsigma = 0.03$	0.14	0.37	0.75	1.0	1.9	2.4

Counterflow Heat Exchanger

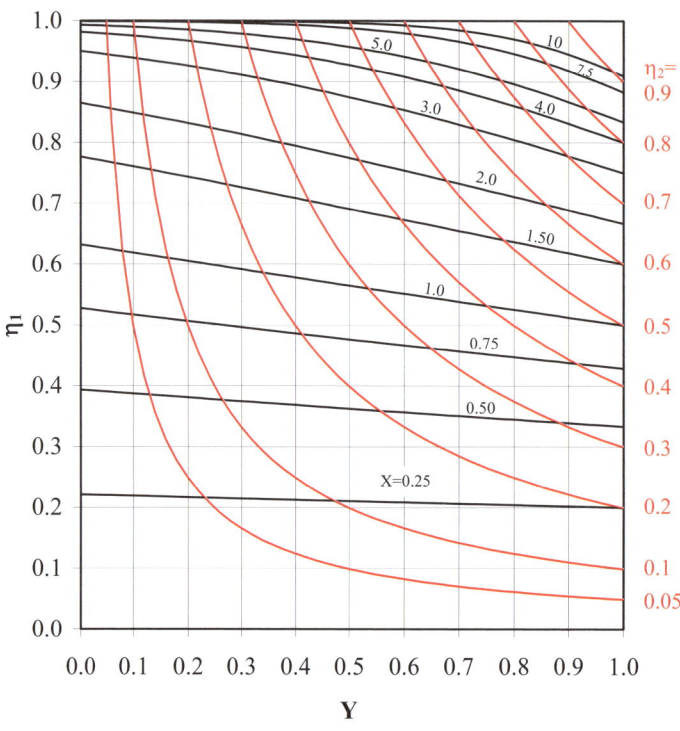

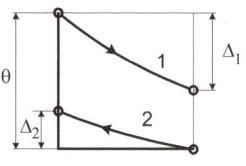

$$\eta_1 = \frac{\Delta_1}{\theta}, \qquad \eta_2 = \frac{\Delta_2}{\theta}$$

$$X = \frac{k \cdot A}{(\dot{m} \cdot c_p)_1}$$

$$Y = \frac{\dot{W}_1}{\dot{W}_2} = \frac{(\dot{m} \cdot c_p)_1}{(\dot{m} \cdot c_p)_2}$$

$$\eta_1 = \frac{1 - e^{-X \cdot (1-Y)}}{1 - Y \cdot e^{-X \cdot (1-Y)}} \quad *$$

$$\eta_2 = \eta_1 \cdot \frac{\dot{W}_1}{\dot{W}_2} = Y \cdot \eta_1$$

* For $Y = 1$, $\eta_1 = \eta_2 = X/(1+X)$

To use diagrams, set
$\dot{W}_1 < \dot{W}_2$.

Parallel-flow Heat Exchanger

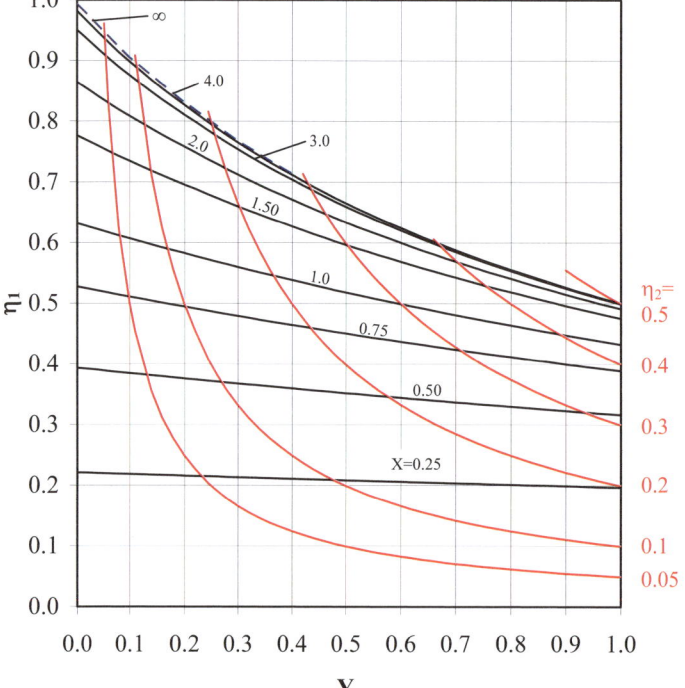

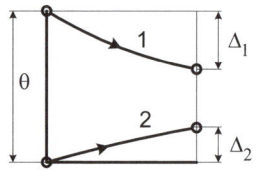

$$\eta_1 = \frac{\Delta_1}{\theta}, \qquad \eta_2 = \frac{\Delta_2}{\theta}$$

$$X = \frac{k \cdot A}{(\dot{m} \cdot c_p)_1}$$

$$Y = \frac{\dot{W}_1}{\dot{W}_2} = \frac{(\dot{m} \cdot c_p)_1}{(\dot{m} \cdot c_p)_2}$$

$$\eta_1 = \frac{1 - e^{-X \cdot (1+Y)}}{1 + Y}$$

$$\eta_2 = \eta_1 \cdot \frac{\dot{W}_1}{\dot{W}_2} = Y \cdot \eta_1$$

Laminar flow heat transfer solutions for fully developed velocity and temperature fields

Geometry		Hydraulic Diameter	Constant Heat Flux	Constant Surface Temperature	Fully Developed Flow Coefficient
		d_h	Nu_H	Nu_T	C
a ◁ 60°		$d_h = \dfrac{a}{\sqrt{3}}$	3.11	2.47	26.6
◯		$d_h = d$	4.36	3.66	32
a ▢ b	$\dfrac{b}{a} = 1$	$d_h = a$	3.61	2.98	28.4
a ▭ b	$\dfrac{b}{a} = 2$	$d_h = \dfrac{4}{3} \cdot a$	4.12	3.39	31.1
a ▭ b	$\dfrac{b}{a} = 3$	$d_h = \dfrac{3}{2} \cdot a$	4.79	3.96	34.2
a ▭ b	$\dfrac{b}{a} = 4$	$d_h = \dfrac{8}{5} \cdot a$	5.33	4.44	36.6
a ▭ b	$\dfrac{b}{a} = 8$	$d_h = \dfrac{16}{9} \cdot a$	6.49	5.6	41
a ▬ b	$\dfrac{b}{a} = \infty$	$d_h = 2 \cdot a$	8.23	7.54	48

Source: Incropera, DeWitt, 1990, **Fundamentals of Heat and Mass Transfer**, Wiley, ISBN 0-471-51729-1

$$f_1 = \frac{C}{Re}$$

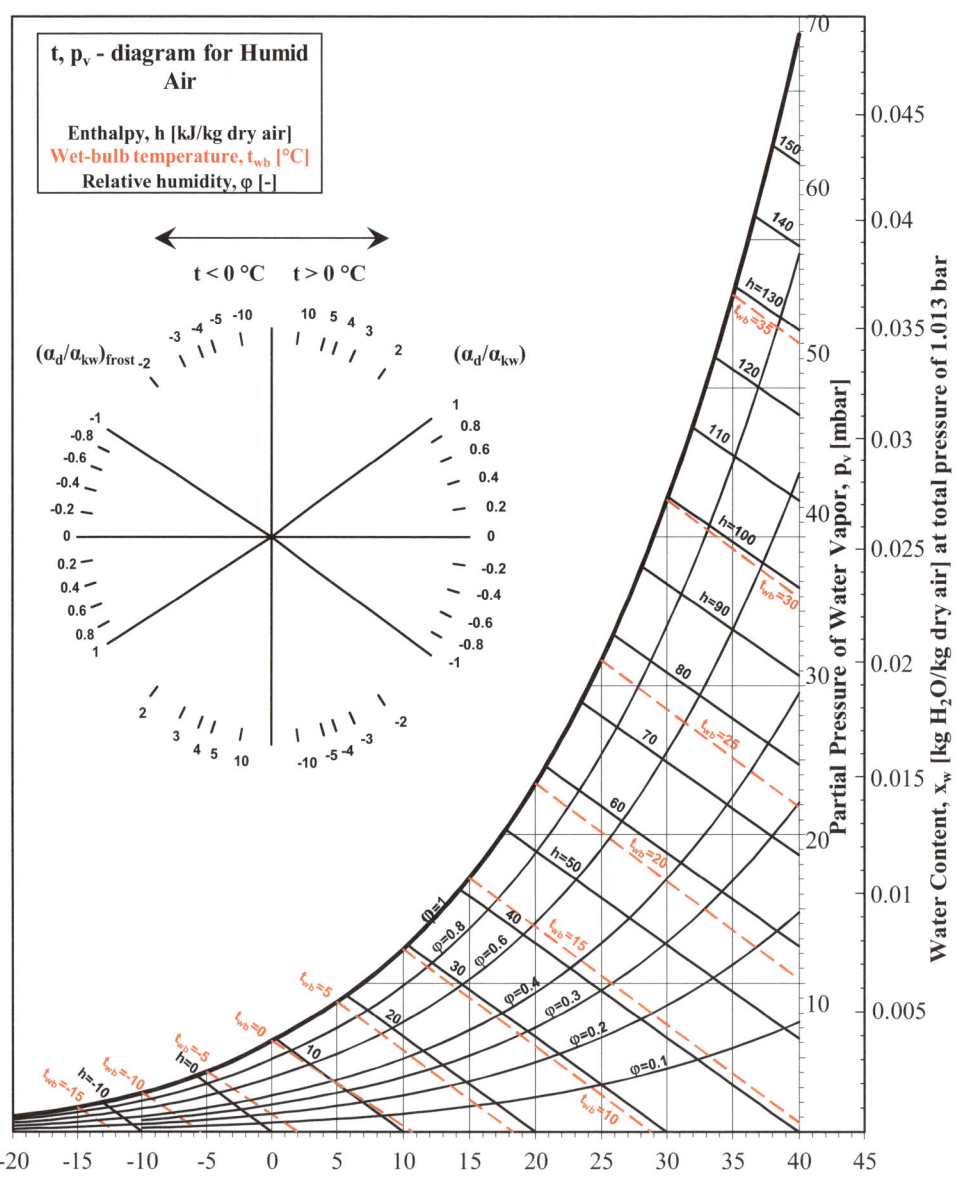

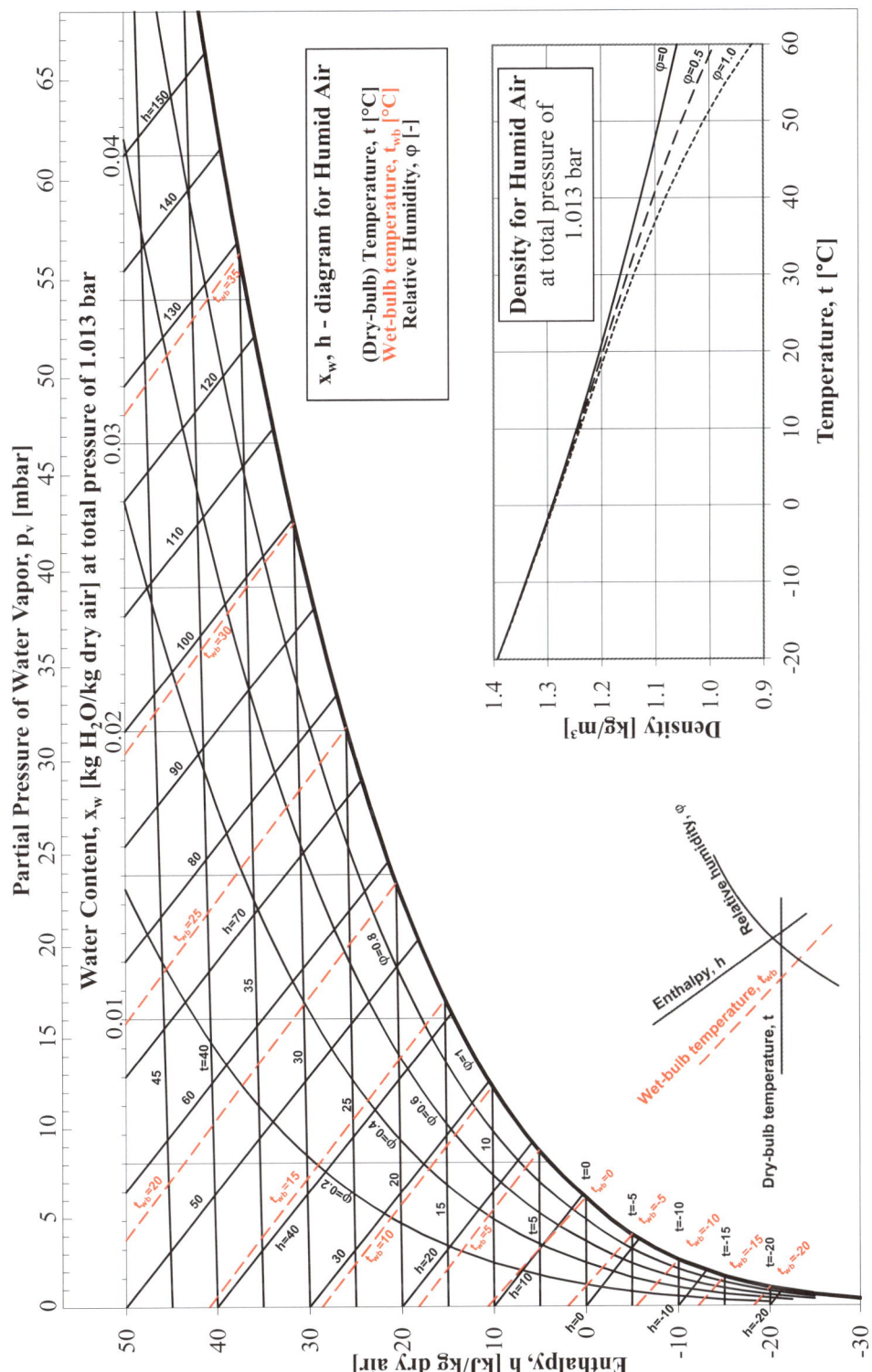

x_w, h – diagram for Humid Air

(Dry-bulb) Temperature, t [°C]
Wet-bulb temperature, t_{wb} [°C]
Relative Humidity, φ [-]

Density for Humid Air
at total pressure of
1.013 bar

Partial Pressure of Water Vapor, p_v [mbar]

Water Content, x_w [kg H$_2$O/kg dry air] at total pressure of 1.013 bar

Enthalpy, h [kJ/kg dry air]

Density [kg/m³]

Temperature, t [°C]

Relative humidity, φ

Enthalpy, h

Wet-bulb temperature, t_{wb}

Dry-bulb temperature, t

Unit conversion

Force

1 kp	= 9.80665 N

Pressure

1 bar	$= 1 \cdot 10^5$ Pa
1 atm = 760 mm Hg	= 1.01325 bar
1 mm Hg	= 133.3 Pa
1 mm H_2O	= 9.81 Pa
1 psi	= 0.0689476 bar

Energy

1 kWh	= 3600 kJ
1 kcal	= 4186.8 J

Power

1 hk (Swedish)	= 735.5 W
1 hp (British)	= 745.7 W

Volume

1 dm³	$= 1$ liter $= 10^{-3}$ m³
1 cm³	$= 10^{-6}$ m³
1 mm³	$= 10^{-9}$ m³

Prefixes

Pico	p	10^{-12}
Nano	n	10^{-9}
Micro	μ	10^{-6}
Milli	m	10^{-3}
Kilo	k	10^{3}
Mega	M	10^{6}
Giga	G	10^{9}
Tera	T	10^{12}

Mathematical Formulas

Integrals

$$\int x^n dx = \frac{x^{(n+1)}}{n+1}, \quad (n \neq -1)$$

$$\int \frac{dx}{x} = \ln|x|$$

$$\int \frac{dx}{x^n} = -\frac{1}{(n-1) \cdot x^{(n-1)}}, \quad (n \neq 1)$$

$$\int (a \cdot x + b)^n dx = \frac{(a \cdot x + b)^{(n+1)}}{a \cdot (n+1)}, \quad (n \neq -1)$$

$$\int \frac{dx}{(a \cdot x + b)^n} = -\frac{1}{a \cdot (n-1) \cdot (a \cdot x + b)^{(n-1)}}, \quad (n \neq 1)$$

$$\int \frac{dx}{(a \cdot x + b)} = \frac{1}{a} \cdot \ln|a \cdot x + b|$$

Logarithms

$$a \cdot \log(x) = \log(x^a)$$

$$\log\left(\frac{1}{x}\right) = -\log(x)$$

$$\log(x) + \log(y) = \log(x \cdot y)$$

$$\log(x) - \log(y) = \log\left(\frac{x}{y}\right)$$